Becoming Spacefarers

BECOMING SPACEFARERS

*Rescuing America's
Space Program*

James A. Vedda

Copyright © 2012 by James A. Vedda.

Library of Congress Control Number:		2012911004
ISBN:	Hardcover	978-1-4771-3092-6
	Softcover	978-1-4771-3091-9
	Ebook	978-1-4771-3093-3

All rights reserved. No part of this book may be reproduced or transmitted in any form or by any means, electronic or mechanical, including photocopying, recording, or by any information storage and retrieval system, without permission in writing from the copyright owner.

This book was printed in the United States of America.

To order additional copies of this book, contact:
Xlibris Corporation
1-888-795-4274
www.Xlibris.com
Orders@Xlibris.com

Contents

Acknowledgements 9

Introduction 11

The Dream

Chapter 1   Are we there yet? 17

Chapter 2   Will we ever get there? 26

The Politics

Chapter 3   Hope, change, and the space program 41

Chapter 4   2010: The year we made conflict 49

Chapter 5   Goodbye space shuttle, hello . . . what? 66

Chapter 6   Transition, or more of the same? 85

Chapter 7   Planes, trains, automobiles, and spaceships 100

The Future

Chapter 8   Finding a path to the mainstream 135

Chapter 9   In search of forward-looking space policy 142

Chapter 10   The Next Great Thing 160

Chapter 11   Answering the big questions...173

Chapter 12   Not the end 187

Acronyms 195

Bibliography 197

DEDICATION

To my late parents, who never understood what I was trying to achieve, but supported me anyway.

Acknowledgements

I would like to thank my colleagues who reviewed the manuscript and provided comments: Marcia Smith, president of Space & Technology Policy Group and editor of SpacePolicyOnline.com; Dr. James Clay Moltz, professor at the Naval Postgraduate School; Dr. Roger Launius, senior curator of space history at the Smithsonian National Air & Space Museum; ubiquitous space journalist Leonard David; and my lifelong friend Dr. David C. Webb, educator, consultant, and former member of the National Commission on Space.

I had discussions with many knowledgeable people in the development of the book. Prominent among them were Dr. Scott Pace, director of the Space Policy Institute at George Washington University; Charles Miller, a commercial space consultant who was a key player in helping NASA Headquarters understand commercial issues; Peter Marquez, vice president for strategy and planning at Orbital Sciences Corporation and former Director for Space on the National Security Council staff; and Dr. Bobby Braun, professor at Georgia Tech and former Chief Technologist at NASA. There were many other conversations, some lengthy and some fleeting, with various associates who stimulated my thinking about this book. Some of them continue to serve the U.S. government and may prefer that their names not be associated with my words. In any case, I thank them all for sharing their insights and providing encouragement.

As always, I'm thankful to my wife Lin, who is so supportive of my work that she even claims to enjoy hearing me talk about it incessantly.

Introduction

Participants in the U.S. space community—people from government agencies, contractors, commercial operators, academia, professional associations, and advocacy organizations—attend a lot of conferences, seminars, and public discussions about the state of space-related professions and projects, and about the outlook for the future. Having participated in a fair number of these gatherings over the past quarter century, it seems to me that there's an unwritten requirement to have at least one speaker at each event say, "We're at a turning point." If that's true, we've been standing at the crossroads for a very long time without making decisions on the nation's space strategy that can endure close scrutiny and the tests of time. If it's not true, a lot of people have been crying wolf for decades, and it will be hard to recognize when we actually do reach a turning point. The objectives of this book are: 1) to make the case that *we really are at a turning point* in space exploration and development, 2) to illustrate what a poor job we're doing at this critical time in our efforts to make strategic decisions on space, and 3) to suggest what the space goals and objectives of the United States should be for the next generation in order to maximize long-term benefits.

The confluence of events and circumstances that defined the state of spaceflight at the beginning of this decade provided visible and compelling evidence that we had reached a critical juncture: the retirement of the space shuttle after 30 years of service; the completion of the International Space Station after nearly three decades of development; the growth of space ambitions and capabilities around the world; and the emergence of an assortment of commercial entities that were willing and (almost) able to take over key functions that used to be dominated by the government, most notably launching humans into space. There's no doubt that the space community has felt its paradigm shifting. Some members of the community are all too confident that they see the path forward when in fact they're looking in the mirror at paradigms past, some of them imaginary. Others are scratching their heads and asking, "What's next?" The final section of this book is my own contribution to this debate, in which I seek to emphasize purpose and lasting value over sprint missions to far-off destinations that pass into history rather than ensure a better future.

It has become a cliché to say that the political challenges are more difficult to overcome than the technical ones in the space business. They certainly seem to take longer because they're more resistant to our efforts to address them. The political forces holding back long-term strategic planning and successful implementation of our space aspirations are short-term thinking, partisanship, and parochialism. To some readers, that may seem like a statement of the incredibly obvious. Maybe so, but that doesn't diminish the importance of tracking and analyzing the specific instances and the mechanisms by which these factors affect what we do in space. This is a necessary part of the search for solutions.

My previous book, *Choice, Not Fate: Shaping a Sustainable Future in the Space Age*, speculated on what we could achieve by mid-century, and included considerable discussion on the need for more and better long-term strategic planning to overcome the many influences that drive us to short-term thinking. That discussion will not be repeated here, although short-term thinking will be in evidence in the events covered in this narrative. Partisanship and parochialism will receive more direct attention, as recent dialogue on the future course of U.S. space exploration and development has reached a point at which partisan posturing and the desire to create local economic benefits have displaced legitimate debate on how to make real progress and serve the national interest.

About six weeks after *Choice, Not Fate* was published in December 2009, President Barack Obama released his budget request for federal agencies for fiscal year 2011. I had no prior knowledge of what would be in the NASA budget request. But I wasn't worried that my book would become instantly out of date since I had focused on a multi-decade perspective rather than near-term programs. My concern at the time was that the president, who had so many high-profile issues on his plate (military conflicts in Iraq and Afghanistan, a fragile economy with high unemployment, energy policy, etc.), would put NASA on cruise control rather than tackle controversial and time-consuming modifications to existing programs, regardless of how necessary they may be. Where would NASA get its guidance at a time when the space shuttle fleet was being retired and the remaining human spaceflight and technology development programs were rudderless and underfunded?

I needn't have worried. To my great joy, the president proposed increasing technology development funding, an important area that had suffered much neglect for at least two decades. Also, his plan set NASA on a path to transition launch operations to the private sector, eventually to include human spaceflight. In the process, NASA's strategic goals would become more capabilities-driven and less destination-driven—another major theme of my previous book.

I was happy, but many others were not. The ensuing battle over contrasting ways of doing business is discussed in some detail in the pages that follow. Combined with circumstances such as a lengthy recession and chronic national debt, the result has been serious questioning of NASA's programs and their priorities, particularly regarding the proper approach to human spaceflight. The delay this has caused is measured in years.

Debate and delay are not necessarily bad things. They could be healthy, even essential for the half-century-old space agency, which must remake itself after completing construction of the International Space Station, retiring the space shuttle fleet, and finding that its increasingly sophisticated science missions are getting increasingly expensive. When it comes to the exploration and development of space, *it's more important to get it right than to do it fast.* In an endeavor this challenging, risky, and expensive, getting it wrong leads to unsustainable efforts, time and resources wasted on dead-end projects, and even tragic loss of life.

The problem with the current debate is that there's no indication it will lead to "getting it right." In fact, the opposite is likely, since the debate is driven by unnecessarily virulent politics and a quest for short-term advantage at the expense of long-term progress. The universe cares little about our beliefs and desires, which can be shattered quickly when confronted by the laws of physics, the state of the art in engineering, and the lessons of history.

Part 1 of this book takes stock of where we are in the quest for the spacefaring dream, from the perspectives of both believers and naysayers. Part 2 assesses where the political debate has taken us in recent years, focusing especially on the distorting influences of partisanship and parochialism. Additionally, the debate over the appropriate roles of the government and private sector remains unresolved and continues to manifest itself in every policy update and budget deliberation. A chapter is devoted to the lessons to be learned from historical analogies, specifically the public-private interactions that enabled the United States to build the infrastructure and regulatory regimes required for maritime, rail, road, and airborne transportation. Part 3 looks ahead, applying lessons from this discussion to search for a space development path that has consistency and purpose. This search leads to the conclusion that the next great thing in space is not human visits to asteroids or Mars but rather the establishment, through the combined efforts of the public and private sectors, of affordable, sustainable space infrastructure in the Earth-Moon system, allowing us to accomplish a broad array of missions that benefit Earth while preparing us to move outward.

Jim Vedda
June 2012

Part 1
The Dream

Chapter 1

Are we there yet?

I'm from Iowa. I only work in outer space.—William Shatner as Captain James T. Kirk in *Star Trek IV: The Voyage Home*

Most people aren't aware that the United States has a statute on the books called the Space Settlement Act, which became law in 1988 and states that "the extension of human life beyond Earth's atmosphere, leading ultimately to the establishment of space settlements, will fulfill the purposes of advancing science, exploration, and development and will enhance the general welfare." In other words, space settlement serves the national interest.

Generations of space futurists and science fiction writers have done their best to convince us that the future of humanity is dependent on expansion into outer space. Their basic message has been as follows: Room to grow throughout the solar system, with its seemingly unlimited material and energy resources, will allow unprecedented levels of wealth generation, improve the human condition, expand the frontiers of science, and give us the means to protect and preserve Earth's ecosystem. And that's even before we move out into the rest of the galaxy.

How are we doing? Clearly, we're not there yet, and we have a long way to go. But that's okay, because this is the mother of all challenges. If we skip steps and try to leapfrog to some desired end state, we will incur massive waste of time and resources, and almost certainly will fail.

Are we at least headed in the right direction and making good progress? That's a harder question to answer, since there is much disagreement about what constitutes the right direction and the right pace. Unfortunately, the indicators aren't good. The most obvious example is our poor performance in following up on Project Apollo. Lunar real estate hasn't seen any increase in property values, or even noticeable development, since humans left their last footprints there in 1972. Ever since then, consensus on how to proceed with the human movement into space—or whether we should even consider such a thing at this point—has been hard to come by. A major difficulty, according to a 2009 editorial in the Cleveland *Plain Dealer*, is "how to sell politicians of constrained vision on the necessity of exploring a limitless universe."

Building space infrastructure—almost

Infrastructure, for most people, is a boring concept. Often it's out of sight and out of mind, like the sewer pipes that run beneath your neighborhood. Even when it's visible, it's usually about as exciting as the pavement on the road. It's also absolutely essential to accomplishing anything big, which will only happen if numerous interdependent components of the infrastructure work properly together—a daunting task in the realm of space operations.

NASA's space shuttle was not an end in itself; it was a piece of space infrastructure. It was envisioned as one of an assortment of tools needed to achieve certain objectives. But what were the shuttle's intended objectives? Not simply to launch routine payloads, which could have been done more cheaply by other means. Satellite deployment missions were used as a justification for investment in the shuttle, but were not a sufficient reason for building it. More importantly, the shuttle would teach us how to build and operate a reusable launch vehicle. It was also needed to carry up the elements and the construction crew for a space station. In the meantime, it could conduct laboratory research in its cargo bay in preparation for work to be done later on the permanent laboratory facilities that the station would provide. So the shuttle was a tool designed mainly to build and service another tool, the space station. The good news is that with these two projects, real space infrastructure was taking shape. The bad news is that the projects took a lot longer and cost a lot more than anticipated, and there was no solid plan for where it was all going.

The space shuttle, despite being just one tool in the space toolbox, was considered by many observers to have been the logical follow-on to Apollo.

That's true in the sense of maintaining continuity in human spaceflight, and keeping the same assortment of NASA centers and contractors actively engaged until the next big space goal came along. But it's not true from the perspective of strategic planning for what used to be called "the conquest of space." In 1969, as the first steps were about to be taken on the Moon, President Richard Nixon assembled the Space Task Group, chaired by Vice President Spiro Agnew, to determine what should come next. NASA presented the group with an ambitious plan designed to move rapidly on several fronts. (The agency's imagination has always been far bigger than its budget.) A space shuttle was just one of the manned space projects recommended, along with an Earth-orbiting station (for 50 to 100 occupants!) by the end of the 1980s, a lunar base and lunar orbiting station by the early 1980s, and a manned mission to Mars concurrent with all of this activity. NASA suggested three possible levels of commitment, but these varied only in the pace of activity, not the content. Then reality struck: the Space Task Group didn't embrace NASA's ambitions, public interest was fading, political support was waning, and budgets were declining. The only human spaceflight program in this grand plan to get the go-ahead was the space shuttle. Approval came in January 1972, and the shuttle made its first flight in April 1981—four years later than expected in NASA's *least* ambitious proposal to the Space Task Group. The International Space Station (ISS) program that we know today, which is much smaller in scope than what was originally envisioned, didn't get presidential approval until 1984 and wasn't completed until 2011—34 years behind the slowest option in NASA's 1969 wish list.

Since Nixon's rejection of the post-Apollo grand plan, we've been building space infrastructure one piece at a time, hoping each one will lead to the next piece, and someday to a critical mass of key capabilities that will allow us to routinely live, work, and establish productive communities in space—in other words, to explore and develop as a spacefaring society. But political and budgetary realities, which have prevented concurrent development of multiple infrastructure elements, have also stretched the timelines of sequentially developed programs so that critical mass can never be achieved. A space infrastructure component is ready to be retired by the time the next one is ready to enter service. The space shuttle was kept around long enough to complete assembly of the ISS, but then terminated just as the station needed its services to support long-term operations. Will the ISS be around long enough to support next steps in on-orbit research and manufacturing, in-space assembly and maintenance, and exploration of other planetary bodies? Unless something changes, it doesn't seem likely.

Although people in some parts of the space community may choose to look the other way as long as funding for their programs is still flowing, this has been recognized as a serious problem at the highest levels. The Augustine Committee, an expert panel appointed by President Obama in 2009 to review human spaceflight plans, found that:

> ...in order to pursue major new programs, existing programs have had to be terminated, sometimes prematurely. Thus, the demise of the Space Shuttle and the birth of "the gap" [in U.S. human spaceflight capability]. Unless recognized and dealt with, this pattern will continue. When the ISS is eventually retired, will NASA have the capability to pursue exploration beyond low-Earth orbit, or will there be still another gap? When a human-rated heavy-lift vehicle is ready, will lunar systems be available? *This is the fundamental conundrum of the NASA budget.* Continuation of the prevailing program execution practices (i.e., high fixed cost and high overhead), together with flat budgets, *virtually guarantees the creation of additional new gaps* in the years ahead. Programs need to be planned, budgeted, and executed so that development and operations can *proceed in a phased, somewhat overlapping manner.* [emphasis added]

This critically important observation in such a high-level study should have become a beacon for guiding the "game changing" space program that the administration sought. Unfortunately, it was overshadowed by the near-term tactical considerations of how to proceed with then-current launcher development programs, to which the Augustine Committee devoted considerable discussion in its report. The paragraph quoted above, in contrast, was buried on page 112.

Are we spacefarers, or not?

Can we call ourselves a spacefaring society today? That depends, of course, on how you define "spacefaring." If it refers to the routine use of space for practical applications like communications, navigation, and remote sensing (including weather monitoring), then most of the societies on Earth are spacefarers because such applications are in widespread use. But that seems too broad, and most people would say that space users are not spacefarers because they're not traveling into space. Several nations have sent a total of a few hundred people into space so far, a tiny elite relative to the world's population. All of them have been launched aboard spacecraft developed by the United States, the Soviet Union/Russia, and China. A more restrictive definition would grant spacefaring status only to those three countries, or only to those individuals who have made a trip to orbit.

Some would suggest that we become spacefaring only when a significant community of humans is permanently living and working off planet, like the familiar but fictional James T. Kirk of Iowa. At that point, key aspects of science fiction scenarios, as depicted in epic series like *Star Trek* and *Babylon 5*, begin to look like a real possibility for the future. But is that future too far away to motivate people today?

There is another approach that could be used to craft a definition. Perhaps it's best to recognize that the definition of spacefaring is a moving target that will change with our growing space capabilities. What we're willing to call "spacefaring" today will not be seen as such in the future. But we can frame the meaning of spacefaring in today's reality, recognizing that it will evolve as we do. For the present era, let's define a spacefaring society as a nation or regional alliance having one or more of the following:

- Operational orbital launch capabilities
- On-orbit human spaceflight programs (for example, participation in the ISS)
- Major facilities and expertise that support space science missions, space launch, spaceflight navigation, and/or spacecraft reentry and landing
- An active space hardware manufacturing sector

This strikes a middle ground that doesn't set the bar impossibly high by current standards (for example, permanent settlements across the solar system with routine interplanetary shuttle service). It doesn't set the bar too low either, so that anyone with a satellite dish antenna would earn designation as a spacefarer. It also avoids the dilemma of identifying only the U.S., Russia, and China as spacefaring nations (because they've demonstrated human spaceflight systems), which would unfairly disparage the important contributions and long-term investments of players like

Canada, Japan, India, and the member countries of the European Space Agency (ESA).

By this definition, many nations, encompassing much of human society, are spacefarers. Go ahead, pause to bask in the glow of that realization: humans have taken the first steps to becoming "citizens of the galaxy" (to borrow from the title of a Robert Heinlein novel). Feels good, doesn't it? Okay, snap out of it, your moment of fulfillment is over. It's time to ask the question: Are we on track to evolve to the next level of spacefaring?

A permanent human presence off the Earth has already begun, but it's unclear when this achievement will grow significantly or whether it will go on without interruption. As of this writing, the ISS has been populated continuously by rotating crews since November 2000, so people are working in space, but not really living there—just visiting for periods of up to six months. And the numbers have been small. Through July 2011, the space shuttle was capable of carrying a crew of seven (although the final flight carried only four), and the largest station crew at the time was six, so the population of low Earth orbit could briefly rise to a double-digit number. (This is in sharp contrast to some predictions in the immediate post-Apollo era, which foresaw an off-world population in the hundreds, or even thousands, by the year 2000.) Not much more can be expected in government programs through 2020, the period that international partners have committed to supporting the ISS, and it's uncertain what will follow. China, India, and other nations have plans for their future occupancy of space, but it's too early to tell when or if these plans will transition to real, lasting programs.

Cultural acceptance of substantial human activity in space, let alone human migration, has been slow. To many, space travel seems like an expensive field trip for scientists or an adventure for thrill seekers. Or maybe an adventurous field trip for thrill-seeking scientists. In any case, it's perceived as a luxury activity, with a lot of resources expended for a privileged few, not for "the rest of us."

In the Apollo era, identification with space was new to most people, and despite the growing presence of satellite communications, weather monitoring, and other space technology influences in their daily lives, space was a news item of passing interest. The Moon landings were certainly exceptional and historic adventures, just as the first expeditions to reach the Earth's poles and the summit of Mount Everest had been a couple of generations earlier. But Apollo flights were just as remote and personally irrelevant as those earlier events because most of us tend to define "adventure" as dangerous things happening in places far away to people we don't know.

Roger Launius, senior curator of space history at the Smithsonian Institution, has tracked U.S. public opinion polls on space going back to the beginning of the Moon race and has shown that at no time during that era, with the possible exception of the period surrounding the Apollo 11 mission, did approval of government spending on lunar trips reach as high as 50 percent. After Apollo 11 achieved President John F. Kennedy's original goal, polls revealed an increasing percentage of respondents—which soon exceeded 50 percent—who thought the country was spending too much on space. So the Apollo program was terminated before all of its planned missions were flown, severely disappointing space professionals and enthusiasts who had devoted a decade of their lives to it and expected to be allowed to continue at a similarly aggressive pace.

Jumping ahead four decades, how much change has there been in public perceptions of the role of space in daily life? Actually, quite a lot when it comes to space applications. In a consumer technology market that features direct-to-home satellite television and broadband services, space-based navigation aids, satellite phones, and overhead imagery on Internet mapping sites, the benefits from space are brought directly into homes, offices, schools, cars, and handheld devices. Importantly in these cases, the space component is much more obvious than in other common applications such as credit card transactions at retail stores and gas pumps, where it's essentially invisible. But today's expanded recognition of space capabilities is still a long way from embracing the building of major space infrastructure and the establishment of working communities there. That's a significant leap that may have to wait for the next generation before it achieves widespread acceptance. Grand visions are realized slowly—or not at all if we're not willing to endure a long process of doing the work and making the investments.

Expanding the playing field

Exactly who will do the work and make the investments? When it comes to developing space for civil and commercial purposes, we tend to think that job falls to our national or regional civil space agency, such as NASA in the United States or ESA in Europe. But a realistic assessment of budgets shows that their contribution, while necessary, is far from sufficient.

NASA's total budget reached a peak in fiscal year 2010 at a little over $18.7 billion. A third of that went to space operations (primarily the shuttle and space station). The 2010 budgets of the rest of the world's civil space agencies combined were nearly equal to NASA's budget. Add it all together and round it up just to be generous, and that yields about $40 billion per year available for civil space spending worldwide. Let's imagine that all of these space agencies get together to build and operate the space infrastructure elements needed to enable an expanding space economy with off-world settlements. And let's say they agree to contribute half of their budgets to the effort, a far higher portion than any of the partners have dedicated to the ISS project. The result is an investment of around $20 billion per year. That sounds like a lot—well over 700,000 times the price of my car! But for the job at hand, it's pitifully inadequate.

To put this in perspective, a comparison to something more familiar is useful. The industrialized nations that are members of the Organization for Economic Cooperation and Development (the OECD, 30 industrialized countries in North America, Europe, and the Pacific Rim, not including China or India) have estimated their future needs for infrastructure investment. For this group of countries, how much must be spent annually to build new capacity and maintain or upgrade existing capital assets? To pick a cheap example, the cost for railways is about $30 billion per year; to pick a not-so-cheap example, roads require about $160 billion per year. The numbers really get scary when you consider the needs of water systems, which are approaching $500 billion per year and are expected to surpass $600 billion by 2025. Electricity grids will need $80-$90 billion per year by 2025, which seems like a bargain compared to water, until you discover that this only includes transmission and distribution, not power generation! The telecommunications sector is harder to estimate with confidence because the technology evolves so quickly, so for now just imagine another really big number.

These seemingly outrageous sums are for the world's most developed nations, so they're weighted heavily toward maintenance and upgrades rather than building new systems from scratch. Development of space will require a lot of "new construction," and we'll need to employ costly launch systems to overcome Earth's gravity and get us to our worksites, which happen to be in an extremely hostile environment. Our $20 billion-per-year contribution from the world's civil space agencies is looking hopelessly puny.

A familiar saying from the early space age warns us: "No bucks, no Buck Rogers." Maybe it's time to update that message: "No commercial contribution, no conquest of the cosmos." Okay, that's not as pithy, but it makes an important point. This can't be done with public-sector funding alone. In fact, in a successful long-term scenario, public funding ultimately will constitute a minority of the investment. That's how the terrestrial infrastructure bills—totaling hundreds of billions of dollars annually for water, electricity, and transportation—are being paid. The global trend in recent decades has been for the private sector to pick up an increasing share of the load. Governments will always have a role due to the societal and national security implications, but non-government entities own, operate, and/or maintain most of the infrastructure. In a 2007 report, the OECD found that:

> ...public budgets fed by taxes will not suffice to bridge the infrastructure gap. What is required is greater recourse to private sector finance, together with greater diversification of public sector revenue sources.

The same will be true as we develop space, an inherently long-term enterprise that can't thrive on government contributions doled out year-by-year, as is done in the United States. But as we'll see in later chapters, not everyone understands the essential role of the private sector or agrees on how or when to make the transition to a more active partnership between the public and private sectors. By way of analogy, Chapter 7 will show that the successful development of transportation infrastructure in the United States was critically dependent on public-private collaboration.

This chapter began by pointing out the existence of the Space Settlement Act of 1988 and quoting the substantive portion of its text. But the original statute was much longer, containing a section with a detailed list of requirements for biennial progress reports that NASA was to prepare for the Congress. What happened to that language is indicative of the priority of space settlement and long-term civil space planning in general. NASA's reports were discontinued in 1994 when the congressional committees that had asked for them told the agency they didn't want them anymore due to lack of relevance to the political environment. In 2000, the reporting requirement was officially stricken from the statute.

The unanswered question in that post-Cold War political environment was whether this signaled a temporary shift in priorities, or a permanent rejection of the dream of humanity's movement into space.

Chapter 2

Will we ever get there?

If it hadn't been for the Cold War, neither Russia nor America would have been sending people into space.—James Lovelock, British scientist

There seems to be near-universal acknowledgement that we'll continue to maintain the space applications that we use routinely today: communications relays, navigation aids, and various types of remote sensing, especially weather monitoring. But there are many people who don't see a future that includes any significant human activity in space, at least not in this century's planning horizon. Their reasoning varies, but generally includes some mix of the following: it's too expensive; it's too risky; it's not relevant to human needs; there is little or nothing to be gained; robotic missions are sufficient to do what we need. In difficult economic times, those who don't see direct and immediate benefits from spaceflight may dismiss it as a useless luxury, soaking up resources that would be better spent elsewhere. Of course, most people lack a vision of a future in space for humans because they simply haven't given it any thought.

Who doesn't like space?

People who are skeptical or indifferent are not just older folks who came of age before Sputnik, or just young people born long after the Apollo era. Nor are they just people who have no connection to the space workforce. Not long ago, a colleague who is a long-time participant in the space community told me in an email that "in economic terms, space is without value and will always be without value . . . not a dime will ever be made beyond GEO," referring to geosynchronous equatorial orbit, the favored location for communications satellites. From the discussion that ensued, the clear implication was that my colleague saw the future of space development as no more than the ongoing refinement of things we're already doing. (For some, that may be sufficient motivation to propel their career, but if I felt that way, I'd be looking for a different line of work.)

A similar view appeared in *The Economist*, a publication that has tended to display skepticism of human spaceflight and adventurous proposals in space commerce. The editors delivered their verdict on this topic in a June 2011 commentary definitively titled "The end of the Space Age"—and to make sure you didn't miss their point, the subtitle was "Inner space is useful. Outer space is history." The commentary reminded readers that the space shuttle was being retired, and stated in no uncertain terms that the ISS will be deorbited in 2020, and at that time "the game will be up." There will be growing economic activity out to GEO, but beyond that "the vacuum will remain empty" because "humanity's dreams of a future beyond that final frontier have, largely, faded." In this view, even robotic deep space missions will face diminishing returns and waning public interest.

Don't expect the private sector to successfully take up the challenge, *The Economist* told us, because the market for commercial human spaceflight is "small and vulnerable" and will get smaller once the ISS is out of service. Nor should we expect emerging space powers like China and India to pick up where we left off, because even if they reach the Moon, they'll stop there, just like the U.S. did, since there's no next step. So it's quite possible, according to *The Economist*, that the GEO altitude of "36,000 km will prove the limit of human ambition. It is equally conceivable that the fantasy-made-reality of human spaceflight will return to fantasy . . . 2011 might, in the history books of the future, be seen as the year when the space cadets' dream finally died."

Pretty harsh stuff, especially if you're one of those space cadets. Note the use of the phrase "finally died" as if to say "it's about time—what took so long?" Essentially, *The Economist* and others of similar persuasion have declared that spaceflight, rather than being the greatest endeavor of all time, is a fad that is quickly fading away.

Such damning commentaries can drive space advocates to strong reactions using descriptors like "ignorant," "short-sighted," "Luddite," and other expressive language that I'll refrain from repeating here. But even extreme punditry can have kernels of truth that should be considered. For a long time, the promise of space has been portrayed optimistically by individuals and groups who have exaggerated the pace of progress, setting themselves up for disappointment and criticism. (For a much more substantial discussion of this, see Chapter 2 of *Choice, Not Fate*.) Significantly, the expectations included not just overly ambitious achievements in the exploration and development of space, but also in societal impacts more generally. For example, author Erik Bergaust, projecting 50 years ahead in 1964, opined: "Who knows, perhaps we will terminate the use of the title *doctor*—because everyone will have at least a Ph.D. degree. That might well become a typical result of our current Space Age brainpower drive." That was over-the-top even for the 1960s, and as we reach that 50-year mark we find that many aspects of our society do not seem to be the products of a brainpower drive.

Nonetheless, there have been profound societal impacts, but so far they have come from a small number of space applications rather than from space exploration and strategic development. Impacts as pervasive and rapid as those brought on by space applications often cause a backlash. Fortunately, there don't seem to be any active organizations in the U.S. or elsewhere whose sole purpose is to oppose space activities. If they're out there, I've seen no evidence of political clout or significant financial backing. However, there are groups that believe some space projects directly or indirectly conflict with their interests, causing them to focus part of their efforts on preventing certain types of activities in space. Some interest groups may choose to lobby against space projects simply because they are competitors for scarce federal funds, but the most popular substantive objections are in response to existing or potential military space operations or the use of nuclear power in space. The dual-use (civilian/military) nature of almost all space technologies fuels suspicion in some communities that all space projects have a military purpose, and may help enlist the efforts of a variety of groups (e.g., environmental, anti-nuclear, arms control) in protests over missions conducted by all sectors of the space community.

Environmental groups in recent times have recognized that space-based capabilities, particularly orbiting sensors of various kinds, are vital tools serving their interests in tracking deforestation, loss of wildlife habitat, air and water pollution, ozone depletion, urban growth, and the manifestations of climate change. But that hasn't always been the case. A colleague once told me of a lunch meeting he had with an environmental lobbyist in the mid-1980s, a time when the movement still harbored distrust of space technology. He told the lobbyist about all the capabilities and benefits of Landsat, the U.S. civil remote sensing satellite system. The lobbyist was intrigued and asked when this great new capability would be available. To his surprise, my friend told him it had been in service for a dozen years (since 1972) and had recently launched the fifth satellite in the series. Environmentalists have grown significantly more space savvy since then.

In another example, environmentalists expressed opposition in the late 1970s to U.S. government studies of solar power satellites that would beam energy down to the terrestrial power grid. This may seem counterintuitive, since solar energy from space was being proposed as an alternative to fossil fuels. However, many environmentalists unwittingly assisted the efforts of the oil industry lobbyists who were trying to kill federal funding for space solar power. The oil lobbyists circulated unsubstantiated information claiming that the satellites' microwave beams would damage the Earth's atmosphere and endanger airplanes, birds, and possibly life on the ground near the beam's receiving antenna. (The potential for these negative side-effects was being studied under government research grants at the time, and is still being investigated.) This episode ended in 1980 when federal funding for solar power satellite research was cancelled, a decision driven by concerns about cost and technological maturity rather than environmental hazards.

Despite the high degree of kinship that has developed between space efforts and environmental issues (for example, in scientific studies and regulatory monitoring), there are still problems. Pollution caused by the manufacture and use of rocket fuel, particularly solid propellants, has been found in farm products such as lettuce. Facilities employed in fuel production activities present an expensive challenge in hazardous waste cleanup for both government agencies and aerospace industry. On top of that, recent studies of possible large increases in the worldwide rate of space launches have found a potential problem in the release of black carbon (soot) into the upper atmosphere. Black carbon in sufficient quantities may exacerbate climate change, and rocket launches inject it directly into the upper atmosphere where it can stay for a long time rather than fall out relatively quickly as it does in the lower atmosphere.

In a different area of interest group concern, an animal rights group, People for the Ethical Treatment of Animals (PETA), has protested the use of animals for experiments in space. At one point the group took its objections to the NASA administrator's office. On October 31, 1996, four PETA representatives protested NASA's involvement in the Russian Bion mission, which was scheduled to carry monkeys into space. In part, the protesters may have been motivated by erroneous press reports that the monkeys would be left to die in space. On its website, PETA later took credit for ending the Bion program and for reductions in other NASA projects that included animal experiments, but it's unclear what influence, if any, the group had on these decisions.

The most widespread and consistent interest group hostility to space activities has been directed against U.S. and allied national security space policy. An example of small but persistent opposition is the Global Network Against Weapons and Nuclear Power in Space, a U.N.-registered non-governmental organization that arranges protests by aligning itself with anti-military, anti-nuclear, and peace activist groups in North America and other parts of the world. In 2006, the group's website criticized the George W. Bush administration's National Space Policy by making linkages between civil and military space programs, interpreting the policy as part of a strategic plan to dominate space and deploy missile defenses and space-based weapon systems. Additionally, the group has attempted without success to prevent the launch of deep space probes with nuclear power sources, such as NASA's Galileo (Jupiter), Cassini (Saturn), and New Horizons (Pluto) missions. An April 20, 2005 press release on the group's website claimed that NASA and the Pentagon had been conducting surveillance and infiltration of the organization, both in the U.S. and Europe. The press release included a threat to sue the U.S. government, although documentation supporting the accusations was not provided and a lawsuit was never initiated.

So far, special interest groups opposed to space activities have been small, fragmented in their approach to space issues, and therefore not influential. For the past two decades, the most serious manifestation of resistance to civil space programs has been the use of the court system by anti-nuclear groups. Opposition to the launch of nuclear-powered space science missions in each case was aimed at the use of up to a few dozen pounds of plutonium to provide modest power levels to spacecraft systems for journeys to the outer planets. Despite strict safety protocols and assessments showing minimal risk, anti-nuclear groups responded with protests and sought the intervention of the courts to prevent launches. To date, the courts have denied requests for restraining orders on NASA planetary launches. However, it cannot be assumed that treatment favorable to the space missions will continue. Only one successful challenge is needed to set a legal precedent, which could have implications for all nuclear-powered missions that follow.

In the future, it's possible that anti-globalization sentiments may become anti-space as well, especially if there is an anti-technology component to the movement. The same entities that dominate space activities—government institutions and large corporations—are seen by critics as orchestrating globalization to serve the wealthy at the expense of the poor. Space technology could be viewed by globalization critics as a tool of transnational corporations that exploit workers, of foreign investors who undermine local businesses, or of wealthy (i.e., spacefaring) countries that economically take advantage of developing nations. So far, space activities have not been directly targeted by globalization opponents, but an anti-technology backlash like the one experienced in the Vietnam era could recur, and space efforts could then become targets.

Reality check for today's space visions

Visions of humanity's future in space have come a long way since the days of Wernher von Braun's spinning space stations and Mars armadas. In the stark reality of modest progress and the realization that costs and risks are even greater than anticipated, there has been plenty of rethinking. But there continues to be failure to articulate a revised vision that is well grounded in not just technology, but also in politics, economics, and history. As we'll see in later chapters, the lessons are there to guide us, but have not been employed adequately to put us on a productive, enduring path.

The futurists and pundits of the early space age weren't the only ones with a tendency toward exaggerated expectations. Too often, today's advocates want to skip essential steps and jump to the end game, as they perceive it. The difference from the early visionaries is that current prognosticators, particularly those old enough to remember the Apollo era, are often driven by frustration at the lack of sufficient progress over the past four decades. This is not entirely a bad thing—discontent can be a useful stimulus. Advocates want to see great strides in space development, particularly ones that they deem important, while they're still active in their careers, or at least while they're still among the living. Perhaps this motivates a desire to do away with intermediate milestones that seem to stand in the way of historic achievement. The risk is that even if we make history one day, the next day we may decide (or be forced) to walk away from follow-on efforts because we neglected those crucial intermediate steps.

A presentation at the annual International Space Development Conference held in Huntsville, Alabama, in May 2011 generated favorable attention and considerable buzz among space enthusiasts on the Internet in the months that followed. The speaker was Jeff Greason, the chief executive of the entrepreneurial launch company XCOR Aerospace, who also had been a member of the Augustine Committee. An engaging speaker, Greason made observations about a topic dear to the hearts of his audience: U.S. space strategy, or the lack thereof. For example, he astutely pointed out that NASA should be designing operational space systems with the intention, from the beginning of the project, that they will be operated by somebody else. But his primary message was a leap too far. Responding to the long-standing complaint that the U.S. doesn't have a goal in space, he countered that there *is* a national goal, and it's settlement. While that may be true of many members of the space advocacy community, there is no evidence that there's anything close to a consensus on this among policy-makers or the general public. In fact, Greason's remarks contradicted the notion of wide acceptance by recognizing that the primary motivation of policy-makers when it comes to space issues is securing government contracts in their districts and preserving the U.S. industrial base.

As noted in Chapter 1, the Space Settlement Act of 1988 was ignored as a guideline for NASA planning, and Congress eventually repealed most of it. The statements of several presidents since the beginning of the space age, as well as congressional language in NASA authorization bills, have made it sound like we're committed to becoming one with the cosmos, but have turned out to be rhetoric rather than resolve.

As far as support for a space settlement goal in a national policy document, recent presidents have come closer. President Obama's National Space Policy of June 2010 directed NASA to "identify potentially resource-rich planetary objects," implying that extraterrestrial resources someday would be exploited for purposes beyond scientific study. (It also directed NASA to "Pursue capabilities . . . to detect, track, catalog, and characterize near-Earth objects to reduce the risk of harm to humans from an unexpected impact on our planet." Although NASA has been cataloging near-Earth objects for a long time, this is the closest any president has come to assigning responsibility to an agency for planetary defense.) President Bush's 2004 space exploration policy endorsed "use of lunar and other space resources," but only in the context of supporting "sustained human space exploration to Mars and other destinations." Aside from that, we've heard no actual declaration of a space settlement goal from any president beyond the usual rhetorical reminders that "it's our destiny."

Greason suggested an approach that would be a hard sell, and displayed an eagerness to skip steps. The goal of his "settlement paradigm" is "permanent and expanding population beyond Earth." The obvious questions are: for what purpose, and why now? His strategy is a bootstrap plan that develops resources at each destination that are needed to reach the next destination. This appears to be based on the old (and, I believe, discredited) destination-driven model, and observers who are not space advocates likely will see it as the ultimate self-licking ice cream cone: Why do humans go into space? So we can go farther into space!

In a February 2010 commentary in *Space News*, humans-to-Mars advocate Robert Zubrin called President Obama's proposal to refocus NASA more on technology development and less on destinations "a horrible mistake that, if accepted, would guarantee zero accomplishment for the U.S. human spaceflight program for the foreseeable future." Zubrin, an outspoken promoter of human spaceflight, believes that NASA should return to what he calls "Apollo mode" in which all efforts are focused on a mission to send humans to a particular solar system destination—Mars being the only sensible choice, in his view. Space technology development that is not connected to a program aimed at a specific destination, he and others have said, would simply create a hobby shop for technologists, producing little that would actually be used.

In fact, "Apollo mode" is precisely the wrong way to go, for reasons that I discussed at length in *Choice, Not Fate*. Project Apollo was designed to accomplish a limited mission as quickly as possible: take astronauts to the lunar surface and bring them back safely. (Yes, there was a science component, but its priority was a distant second.) Once the primary mission was completed, there was no accepted and funded plan for building on that accomplishment, despite the dreams of many. NASA had brilliantly achieved a national goal, both physically (reaching the Moon) and psychologically (winning hearts and minds around the world). The United States earned bragging rights of historic proportions that no one could take away. And then we put most of the remaining flight hardware in museums. It had been designed for a very narrow mission, and we weren't willing to spend what was needed to adapt it to new missions, other than the short-lived Skylab space station and the one-shot Apollo-Soyuz linkup, both of which were completed less than three years after the last Apollo lunar flight.

It's unrealistic to believe that NASA can or should go back to the way it was in the 1960s. "Apollo mode" was unsustainable because it was *designed* that way—the agency spent an entire decade on an adrenalin high. At the program's peak, NASA employed twice as many civil servants as it has in recent years, plus it supported approximately 400,000 contractor personnel. Many of its 1960s-era technology development projects picked the low-hanging fruit of the early space age, when the demands were simpler and failure was more readily accepted as a cost of doing business in a new area. And of course, there was the funding advantage. Some, including Zubrin, have tried to dismiss this, but it doesn't require advanced math to see how important this was. Adjusted for inflation, and averaged across the Apollo era, the space agency had at least nine percent and possibly as much as 27 percent more buying power than it had in 2010, its peak funding year in recent history. No one expects to see increases of that magnitude any time in the foreseeable future. (There are multiple inflation indexes, so I chose the smallest and largest ones I could find to provide a range. I multiplied the total NASA budget for each year from 1961-1972 by the inflation factor for that year to convert it to 2010 dollars. The inflation-adjusted yearly average for these 12 years fell in a range of $20.4 billion to $23.8 billion, compared to the actual NASA budget for fiscal year 2010 of $18.7 billion.)

President Obama, his science advisor John Holdren, and his NASA Administrator Charlie Bolden all have referred to Mars as "the ultimate destination." Similar remarks have come from President George Bush (both Sr. and Jr.) and other commentators inside and outside the space community. Those who favor going back to the "Apollo mode" almost invariably have their sights set on Mars as the next target for humans. For decades, there has been an assumption that once we've reached the Moon, the logical next step is to turn our attention to the red planet.

A human mission to Mars seems to be firmly fixed in the American space psyche. Now here's the part that some will find heretical: It shouldn't be. I don't believe Mars is "the ultimate destination" in space—when we reach it, we're not going to stop looking beyond it—and I question whether such a label makes sense in any case. At our current stage of development, we're not sophisticated enough to be talking in such terms.

The logic for sending people to Mars as the next great space endeavor is extremely weak. The idea is rooted in science fiction and in the space exploration scheme popularized by rocket engineer Wernher von Braun and his colleagues in the early 1950s. Proponents remind us that Mars is the most Earth-like body in the solar system, but that's not saying much. Sure, it has a solid surface and a day/night cycle close to ours (about twenty-four and a half hours), but it's cold, devoid of any kind of life we've been able to find so far (even microbes), and its atmosphere is so thin that it would be indistinguishable from a vacuum to any human trying to breathe.

If the intent is to explore Mars and maintain a permanent presence there, infrastructure requirements would include habitable facilities at multiple locations; surface transportation far more capable than the old lunar rovers; constellations of satellites to provide Mars with communications, navigation, and routine overhead surveillance; and the ability to produce most consumables locally, including air, water, food, and energy. That's a serious long-term commitment of resources. Even if funding was not a problem, there's still a lot we don't know about how to do these things on a big enough scale, and with sufficient reliability, at such a great distance. Obviously, this is a "giant leap for mankind" far beyond the one we achieved in the Apollo program, which involved little more in off-world operations than the round-trip transportation for a jaunt lasting substantially less than two weeks.

Much to the chagrin of many Mars enthusiasts, this is NOT what near-Earth space looks like. Mars is not a second, slightly farther moon of Earth. In fact, its distance from Earth ranges between 100 and 1,000 times greater than our Moon.

Mars is a superb target for scientific investigation, but how much are we willing to invest and risk at this stage of our development to have it done by humans? Mars may be just the next planet out from the Sun, but at its *closest* point, it's about 145 times farther than the Moon. The greatest separation between Earth and Mars—when the two planets are on opposite sides of the Sun—stretches to over 1,000 times the Earth-Moon distance. That's one heck of a long logistics chain. (Too bad Venus isn't a good choice. It's only 99 times farther than the Moon at its closest point, and it's near in size to the Earth. Unfortunately, its dense atmosphere is quite toxic and the temperature all over the planet gets hot enough to melt lead.)

Hastily jumping off to other planets, even the closest ones, skips important steps. If we want humans to move out beyond Earth in a productive and sustainable way, we have to learn to live, work, build, and produce in the local neighborhood first. The technologies and operational experience needed for venturing to Mars haven't been demonstrated yet, even right here in the Earth-Moon system.

Legendary aerospace engineer and futurist Krafft Ehricke purportedly said, "If God had wanted man to become a spacefaring species, He would have given man a moon." Well, God did give us a moon, so the obvious implication is that He does want us to go into space. But He could have given us a little more encouragement to lead us out to the rest of the solar system: several moons to draw us out a little farther each time.

Another handy thing about having a moon to practice on, in addition to being relatively close, is that it maintains about the same distance from Earth, giving us plenty of convenient launch windows for both going out and coming back. Other planetary bodies in the solar system, to our great and lasting inconvenience, are nowhere near as considerate. They travel in their own heliocentric orbits that cause extreme variations in their distance from Earth, which severely limits launch opportunities using efficient trajectories. If Earth had multiple moons at incrementally increasing distances, that would provide a rich training ground to prepare us for the much longer leap to the rest of the solar system. But alas, we have just one moon, forcing us to obtain our training in this limited sphere of only a half-million miles in diameter called cislunar space. From here, we must decide when we're ready to take on a challenge that is exponentially greater.

The reasons to go

The rationale for a capabilities-driven approach to exploration and development of the solar system can be found throughout human history. What has motivated human societies, usually at great cost and risk, to undertake major migrations to, or activities in, unfamiliar and challenging environments? It comes down to two things. First, they go where the resources are. Humans in search of precious minerals, raw materials, and energy, and the wealth they bring, have explored the most hazardous environments on Earth, including the ocean floor, the polar regions, treacherous terrain, and underground mines. Valuable discoveries have spawned economic booms and determined human migration and settlement patterns. Second, they search for new avenues to solve their problems and improve their living conditions. There are many examples of communities of people moving to escape a deteriorating environment (famine, drought, overcrowding, etc.), political or religious persecution, and other conditions that didn't allow them to grow, or even eke out a sustainable living.

This familiar storyline will play out again as humanity gets busy beyond Earth. That's because:

- **Space is where the resources are.** The resources of the solar system are abundant beyond our foreseeable ability to fully exploit them. Although hostile, the space environment offers potentially useful properties such as microgravity, vacuum, continuous solar energy, and isolation from Earth (which can be beneficial for hazardous activities). Solid bodies offer materials that we can learn to extract, process, and use, possibly in place of resources extracted on Earth.
- **Space will provide new avenues to help us solve problems, improve the human condition, and ensure the continuation of our species.** The initial movement into space to look for solutions started decades ago, with more machines than people. The problems we've sought to address have included political and military tensions between nations, the need for faster and more comprehensive communications, mitigation of the destructive forces of nature, and measurement of the effects of our actions on the health of the planet. These efforts will persist, and the tools and facilities will grow in size, number, and capabilities, drawing sustenance from the extraterrestrial energy and material resources that will be developed in parallel. New problem-solving missions will become feasible,

such as moving environmentally damaging activities off the Earth, or averting the destruction of an incoming asteroid.

A path to realizing this future will be explored in the final section of the book. Before considering that path, it's important to recognize that naysayers and protesters are not the only threats and are unlikely to be the worst threats facing space aspirations. Those are most likely to come from two sources, both inherent to large, complex endeavors. The first is economics. High costs and long lead times are well known characteristics of the space business, as they have been for all endeavors that require massive inputs before the first revenue-generating outputs are produced. The second is politics, where short-term thinking is coupled with partisanship and parochialism, resulting in a suboptimal decision environment (to put it kindly) like the one described in the next section.

Part 2
The Politics

Chapter 3

Hope, change, and the space program

> *We are still a leader in space exploration, but frankly I have been pushing NASA to revamp its vision . . . Now what we need is that next technological breakthrough. We're still using the same models for space travel that we used for the Apollo program . . . Let's allow the private sector to get in so that they can, for example, send these low-Earth-orbit vehicles into space.*
> —President Barack Obama, July 2011

In the space community, we like to think of ourselves as pretty important. How could anyone not see the lasting value of what we're doing? But in the grand scheme of things, far bigger issues compete for attention and funding, especially election-drivers like the economy, overseas conflict, and big entitlements such as Social Security and Medicare. In that environment, space issues seem like a sideshow.

When Barack Obama became president in January 2009, the number and urgency of issues on his plate seemed to eliminate any possibility that the civil space program would garner high-level attention. Military conflict continued in Iraq and Afghanistan. Terrorism still appeared to present a serious threat. International relations had deteriorated in recent years. Most of all, the U.S. economy was teetering on the brink, threatening a reenactment of the Great Depression of the 1930s. Add to that all the other chronic problems like education, health care, and climate change. Anyone running for President of the United States had better love a challenge—or more accurately, a whole pile of them.

With his job jar filled to overflowing on day one, the logical assumption would have been that space programs would be far below the president's radar for a long time to come. But miraculously, there was swift action. Within four months, Obama had signed a study directive that would lead to a new National Space Policy the following year, and had set up the high-profile Augustine Committee mentioned in Chapter 1, which would influence an overhaul of human spaceflight a few months later.

At first glance, it might seem that the Obama administration had great momentum already in place for civil space programs thanks to its predecessor. The so-called "Vision for Space Exploration" aimed to return humans to the Moon by 2020 and target Mars after that. Implementation of that plan got underway with the startup of NASA's Constellation program in 2005. Constellation encompassed the development of two new rockets for crew (Ares 1) and cargo (Ares 5), a crew capsule (Orion), and a lunar lander (Altair).

Closer examination reveals that the program had been underfunded from the beginning. Greater-than-expected costs of returning the shuttle to flight after the 2003 *Columbia* accident contributed to the shortfall. But more significant was the Bush administration's failure to request the billion-dollar increase in NASA's budget that was originally promised. That increase was supposed to carry the program through its first five years, after which the shuttle fleet would be retired and Constellation, it was presumed, would absorb the shuttle's $3 billion per year budget. This was a questionable assumption, since a future president and Congress would have to agree that this sizable appropriation would simply shift to another program within the same agency—hardly a guaranteed outcome in a difficult fiscal environment.

If the program was to continue on schedule, funding would need to ramp up enormously in the new administration. Substantive work on Ares 5 and Altair had not begun by the end of 2008. Ares 1 and Orion had progressed to hardware development, but neither was ready to conduct a test flight by the time Bush left office. A multi-year gap between the last space shuttle mission and the first flight of the Ares 1/Orion had been preordained in 2004, and the gap had widened to as much as seven years by 2009. As long as the gap continued, the U.S. would have to buy seats on Russian Soyuz capsules to maintain access to the International Space Station, at a cost of over $51 million each, rising to $63 million in 2015.

But the most troubling issue that was left behind was the workforce disruption that would occur when the space shuttle program ended. The Bush administration could not have known in 2004 that unemployment would be the nation's biggest concern in the years following 2008, but it was obvious from the beginning that the civil space workforce would be hit hard by the absence of shuttle flights after 2010. Thousands of skilled government and contractor workers would be displaced, perhaps more than 8,000 around Florida's Space Coast alone, and an equivalent number around the rest of the country.

Publicly, the Bush administration claimed that the majority of shuttle workers could be retained during the gap by transferring them to Constellation and other NASA programs. What they failed to say is that this referred to *government* workers, not contractors. An internal NASA workforce strategy team developed options for holding on to civil service expertise through retraining, relocations, and other means. By 2007, plans for retaining up to 90 percent of the 2,000 threatened civil service jobs had been crafted. And Congress was watching: starting in 2008 and continuing "until the successor human-rated space transport vehicle is fully operational," NASA was required to deliver a semi-annual workforce transition strategy to its oversight committees.

In contrast, saving a significant portion of contractor jobs was never realistic, for two reasons. First, the contractor personnel who processed shuttle components and payloads and prepared launch pads were not the same ones who would design and test new spacecraft and rockets, and they were not necessarily in the right locations. NASA could not afford to carry thousands of contractor employees on the payroll who would have little to do until launches resumed in several years. Some fraction of the workforce might be eligible for reassignment but the majority would still be out of work. The second reason that job loss was inevitable is that one of the primary rationales for replacing the shuttle with Ares rockets was to make launch operations more affordable. It doesn't take mathematical wizardry or advanced business acumen to realize that you won't save money if the same size workforce is retained at the same wages. Ares launch operations would need to use a smaller workforce than shuttle operations, or else there would be no significant savings.

This would have been an unpleasant surprise to leave for the next administration in any case, but its importance was compounded by the general problem of rising unemployment that greeted the incoming Obama team. It became one more ugly situation that could have future electoral consequences, especially in the swing state of Florida.

In December 2008, during Obama's post-election transition period, a team at NASA Headquarters considered some what-if scenarios that focused on possible paths that major programs could take under the new administration. One scenario included cancellation of the Ares rockets and lunar lander. At the time, this was seen as a worst-case situation that would unravel the agency's recovery plan for civil service jobs. This is exactly the situation NASA would face about a year later.

The nation's aerospace workforce had been troubled for over a decade, its average age increasing due to the lack of sufficient fresh blood entering the industry. The Bush administration expressed confidence that young talent would flock to aerospace jobs, as they had in the post-Sputnik era, as a direct result of its Moon-Mars plan. Clearly, this didn't happen. In the six years between Bush's announcement of the plan and Obama's move to cancel the Constellation program, there was no improvement in the robustness of the aerospace workforce. Indeed, the indicators continued to decline. Nor did Obama's changes or a congressionally mandated launcher project turn things around. In all cases, the potential workforce responded very differently than the Apollo generation. Instead of trusting that the government would follow through on its ambitions, their behavior suggested a wait-and-see attitude: "Show me the money!" So far, the money has not been sufficient to prompt technically savvy young workers, who have far more options than previous generations, to set their sights on the aerospace industry.

The Augustine Committee: a window of opportunity

An editorial in *Aviation Week & Space Technology* signaled the magazine's approval of the choice of former Lockheed Martin CEO Norm Augustine to lead Obama's committee of experts (formally called the Review of Human Spaceflight Plans Committee), and expressed confidence that Bush's Moon-Mars plan would be reaffirmed. The editorial concluded that "the prudence reflected by a three-month study may serve [Obama] and the country well."

The Augustine Committee's objective, as stated in the group's charter, was to review ongoing programs

. . . to ensure the Nation is pursuing the best trajectory for the future of human spaceflight—one that is safe, innovative, affordable, and sustainable. The Committee should aim to identify and characterize a range of options that spans the reasonable possibilities for continuation of U.S. human spaceflight activities beyond retirement of the Space Shuttle.

From June to October 2009, the Committee held several meetings, many of them open to the public. (I attended the three public meetings that were held in Washington, DC.) They received input from a wide variety of stakeholders: NASA program officials, aerospace contractors, space scientists, members of Congress, former government officials, foreign space agencies, and space advocates. Each made their case for either continuing or altering the Constellation program. For example, NASA managers portrayed Ares/Orion development as being on track for its first manned flight in 2015, with its costs under control. In contrast, United Launch Alliance, a Boeing/Lockheed Martin joint venture that operates the Atlas and Delta rockets for U.S. government customers, asked the panel to consider its Delta 4 Heavy rocket as an alternative to Ares 1, claiming it would be safer, cheaper, and would be ready to carry crews by 2014. Others suggested using nascent commercial vehicles or simply starting over with a new design.

Veteran aerospace industry executive Norm Augustine, chairman of the Review of Human Spaceflight Plans Committee, confers with physicist, educator, and former astronaut Sally Ride, one of the committee's members, at an August 2009 public meeting. Source: NASA

Ultimately, the Committee found that the Constellation program was in good shape technically and managerially, but its budget was problematic. The resources it had been receiving were inadequate, and the situation was only going to get worse unless something changed dramatically. The Committee took a step back and looked at the problem from a destination perspective: What should be next? They came up with three possibilities that they called Moon First, Mars First, and Flexible Path. The first two are self-explanatory; the third refers to an assortment of objectives such as lunar orbit, near-Earth asteroids, the moons of Mars, and gravitationally stable locations in space called Lagrange points. Mars First was eliminated as an inappropriate near-term goal due to deficiencies in key capabilities and anticipated budget constraints.

The Committee's report discussed several possible architectures and timetables to execute Moon First and Flexible Path. The Obama administration chose the latter approach, although a clear rationale for that choice was never publicly articulated. People close to the decision process have indicated that it was at least partially motivated by a desire to advocate something different from the Bush plan. But the biggest driver may simply have been concerns about both near-term and long-term budgets, which made the "flexible" part of Flexible Path look more attractive, to the detriment of Obama's otherwise enthusiastic views about America's future in space. Presidential candidate Obama in August 2008 stated that "the United States should maintain its international leadership in space while at the same time inspiring a new generation of Americans to dream beyond the horizon." He promised to "establish a robust and balanced civilian space program" that would "inspire the world with both human and robotic space exploration" while at the same time "confronting the challenges we face here on Earth, including global climate change, energy independence, and aeronautics research." He believed "a revitalized NASA can help America maintain its innovation edge and contribute to American economic growth." After receiving the results of the Augustine Committee report in late 2009, the president's advisors hit him with the bitter reality of the deteriorating outlook for budget deficits, telling him—reportedly in bold, underlined type—"Especially in light of our new fiscal context, it is not possible to achieve the inspiring space program goals discussed during the campaign."

Responses to the Augustine Committee report from Capitol Hill were swift and generally not favorable. For example, Senator Richard Shelby, an Alabama Republican, sought to discredit the report's suggestion that more launch responsibilities, including the launch of astronauts, could be shifted to commercial operators: "Without an honest and thorough examination of the safety and reliability aspects of the various [commercial] designs and options, the findings of this report are worthless . . . Pretty slides and unproven promises will not show us you have the right stuff to be entrusted with the lives of our astronauts." He acknowledged that someday, private companies may be ready to take over launches to low Earth orbit, but "That day is not today and it will not be for years to come."

House Republican Bill Posey of Florida reacted to the report's release with remarks that attempted to create the false impression that underfunding of Constellation began with the newly arrived Obama team. He charged that "the budget proposed by the Administration falls far short of making NASA a priority... Is it any wonder that Constellation is not meeting its schedule when you underfund it to the tune of $20 billion?" The congressman went on to argue that shuttle flights should be continued through 2015, as specified in legislation he introduced in April 2009—too late to be implemented without unacceptably high costs.

Among Democrats, Representative Parker Griffith of Alabama said that the report "lacks the ambition and drive that first put our astronauts in space, beat the Russians to the moon, and is synonymous with the American space program... These findings are incompatible with our national goals to return to the Moon, Mars and beyond, and we in Congress will not stand for it. We can do better." Sen. Jay Rockefeller of West Virginia took a completely different tack, questioning the justification for human spaceflight by stating that "this is no longer the era of Apollo and the Cold War where the payoffs for advancing the space and Moon agenda are unquestionable."

In general, members of Congress thought that the way ahead for space exploration had been settled in a bipartisan manner and enshrined in the NASA Authorization Act of 2008. Just a year later, the new administration was trying to reopen what they saw as a done deal. In their rhetoric, the members complained that making changes would undermine U.S. efforts to maintain its leadership in space. Less talked about by the members, but undoubtedly more important to them, major changes would disrupt the distribution of funding and jobs to key locations around the country. Their brewing discontent was a foreshadowing of what was to come in 2010.

Chapter 4

2010: The year we made conflict

You are entitled to your own opinion but you are not entitled to your own facts.—Senator Daniel Patrick Moynihan

An annual ritual took place in Washington on February 1, 2010. As usual, it was a big deal "inside the Beltway" but drew scant attention around the rest of the country, other than the obligatory mention of its occurrence in the daily news. The president delivered his budget proposal for the next fiscal year (FY 2011) to the Congress.

Each year, this event entails more than just transmitting, from the White House to Capitol Hill, a thick document containing an excruciating level of detail on agency budgets that can be appreciated only by hard-core budget-followers. Every agency in town stages a press conference for the benefit of the cadre of journalists who monitor its activities. Across the spectrum of agencies, the questions are very similar: Does the proposal include any major new starts or program cancellations? Are there any significant increases or decreases to existing programs? What does all this say about the president's priorities?

Immediately after the release of the annual document, the customary press releases issue forth from members of Congress. Some of the president's partisans will say things like, "We'll work with the White House to pass this budget." Members of the opposition party will declare it "dead on arrival." Members of both parties will pick out a favorite program and praise or lament the funding level recommended by the president.

President Obama's FY11 proposal for NASA quickly stirred up a storm of protest from Capitol Hill. Lawmakers on both sides of the aisle called the proposal "radical." It's no coincidence that the most vocal of the unhappy legislators represented areas of the country that stood to lose contracts and jobs if NASA changed course in the manner the president was suggesting. Some expressed their objections in nearly apocalyptic terms.

What did the president propose that upset so many people? Here's a summary of the high points, derived from the White House Office of Management and Budget (OMB) documentation released in conjunction with the budget rollout:

- Add $6 billion to NASA's budget over the next five years for the human space exploration program.
- Initiate development and demonstration programs of "game-changing" technologies that will increase the reach and reduce the costs of future human space exploration and other space activities.
- Contract with American companies to provide astronaut transportation to the ISS, reducing dependence on foreign crew transport capabilities.
- End the Constellation program, which was developing launchers and other systems to enable a return to the Moon by 2020 using an approach similar to the Apollo program.
- Extend ISS operations to at least 2020 and increase its utilization.
- Enhance global climate change research and monitoring.
- Provide for a robust program of robotic solar system exploration and new astronomical observatories.
- Revitalize and realign NASA as an efficient 21st century research and development agency.

Overall, this looks very positive for the space agency and seems to be a forward-looking strategy, not one that would provoke hostility. But it was as if the president had thrown down a gauntlet and said, "Let the games begin."

Negative reactions

The trade press responses to the president's proposal ranged from cautiously hopeful to caustically skeptical. *Space News*, the space community's weekly newspaper, saw the changes as doomed to failure, editorializing that there was "cause for serious concern" and concluding that the president's plan "so far has the look, feel, and smell of a dead-end vision that maximizes the risk and minimizes the potential reward of human spaceflight." *Aviation Week & Space Technology* had a more nuanced view, stating that "we have no quarrel with the ultimate objective of easing NASA out of the role of central planner for all U.S. human activities in space" and recognizing that if the plan were to pay off, "history may record this as a turning point in the evolution of spaceflight." However, the magazine withheld its endorsement, asserting that "NASA will have to add substantial details to this sketchy policy shift" which "strikes us as 'redirection' without a compass."

As would be expected, the U.S. aerospace industry had mixed views about the wisdom of the move, depending on whether they saw it benefiting or hurting their own interests. Orbital Sciences Corp. and Space Exploration Technologies Corp. (SpaceX), both of which were already receiving NASA support to develop commercial launch services, clearly saw their fortunes rise with the potential for taking over responsibilities that traditionally have resided in the government. On the other hand, Lockheed Martin (under contract to develop the Orion crew capsule) and Alliant Techsystems (also known as ATK, maker of solid fuel boosters for the cancelled program's launch vehicles) were looking at big losses in contract dollars and in their workforce.

Alabama is home to the NASA Marshall Space Flight Center, where rockets are designed and their engines are tested. The state's two Republican senators had a very dark view of the proposal. Sen. Jeff Sessions felt that it "abandons our nation's nearly five-decade commitment to human spaceflight and will likely result in NASA taking a back seat to China, Russia, and India in space exploration." His colleague Sen. Richard Shelby called it "the death march for U.S. human spaceflight." Shelby's indignation only increased in the weeks that followed. At an April 22 hearing that featured testimony from NASA Administrator Charles Bolden, he delivered a statement designed to heap scorn on Bolden, the Obama administration, and the emerging commercial spaceflight sector. He judged that the president's plan

> ... abandons our nation's only chance to remain the leader in space and instead chooses to set up a welfare program for the commercial space industry ... the Administration claims that if we build up this so-called commercial

rocket industry the private sector market will magically materialize to produce more expendable launches, at a lower cost, earlier than the schedule of Constellation . . . This request represents nothing more than a commercially-led, faith-based space program. Today, the commercial providers that NASA has contracted with cannot even carry the trash back from the space station much less carry humans to or from space safely . . .

Texans had something to say about this as well, since the Johnson Space Center in Houston houses astronaut training, spacecraft design, and Mission Control. Rep. Pete Olson questioned Obama's intention to sustain the space program, and fellow Republican Ralph Hall was "alarmed that the administration is planning to reject all that we have accomplished."

Also greatly concerned was the Florida delegation, with its large constituency of space workers in the area around the Kennedy Space Center often referred to as the Space Coast. Republican Rep. Bill Posey called Obama's plan "a giant step backwards." The vehemence of Republicans could have been expected, since the proposal for change was coming from a president of the opposing party. But Democratic ire was also clearly evident, though less inclined to disparage the president. After all, they too were facing the likelihood of jobs and federal contracts moving out of their districts. Among Democrats, Rep. Sheila Jackson Lee of Texas said "we're going to be very vocal about any undermining of the commitment to NASA," and Rep. Suzanne Kosmas, from the Space Coast, called the proposal "unacceptable."

The negative responses were not limited to members from districts with NASA field centers. Rep. Bart Gordon, a Democrat from Tennessee and chairman of the House Science Committee, felt that "The space agency's budget request represents a radical departure from the bipartisan consensus achieved by Congress in successive authorizations over the past five years. This requires deliberate scrutiny . . . We will need to hear the administration's rationale for such a change and assess its impact on U.S. leadership in space before Congress renders its judgment on the proposals." Rep. Frank Wolf, a Republican who represents part of Washington's Virginia suburbs, leaped to some extreme conclusions in a *Space News* commentary: "I believe it is unacceptable that the administration would gamble our entire space exploration program for the next five years on research. The dirty little secret of this budget proposal is that it all but ensures that the United States will not have an exploration system for at least two decades." Wolf also expressed displeasure at what he saw as the administration's arrogance in the way it controlled the decision process. He even complained about the "smug facial expressions" of White House staffers sitting behind presidential science advisor John Holdren as he testified at a hearing.

Some of the most vocal objectors chose to put their own spin on the facts to make their case. In a *Houston Chronicle* commentary, Texas Republican Sen. Kay Bailey Hutchison complained that "If President Obama has his way, the U.S. will retire the space shuttle program later this year, just as the International Space Station is finally complete and without a viable alternative to take its place." From that statement, one might think that it was Obama who initiated the shuttle's retirement and the ensuing hiatus in U.S. human spaceflight. In fact, it was President Bush who took that action in 2004, initially proposing a four-year gap that grew to at least six years by the time he left office. It was too late to close the Bush gap by the time Obama took office, but misstatements about his role prompted Obama to remind everyone in an April 2010 speech that the gap was "based on a decision that was made six years ago, not six months ago."

Ironically, in her commentary, Sen. Hutchison went on to describe legislation she introduced to keep the shuttle flying until a replacement was ready. That solution had little chance of passing in 2010 and was too late to be implemented even if it did. Apparently, she was not motivated to suggest similar legislation (and contradict a president of her own party) when the problem first arose years earlier, and there was a chance that it could have made a difference. The NASA authorization bill passed in 2005 said nothing about continuing shuttle flights to eliminate a human spaceflight gap—instead, it simply stated that it was U.S. policy to have continuous human access to space, recognized that this wouldn't happen, and requested reports from NASA on how the agency would handle the gap.

Another example of creative writing came from the Republicans of Utah's congressional delegation, Senators Orrin Hatch and Bob Bennett, and Representatives Rob Bishop and Jason Chaffetz, none of whom were on space-related committees or were otherwise routinely involved in space issues. What they had in common was that booster-maker ATK, a big contractor in the Ares rocket project that was threatened by the new proposal, was located in their state. That made them staunch defenders of "the Ares 1 rocket, which has come so close to creating an infrastructure for traveling to Mars," as they described it in an April 2010 commentary in *Aviation Week & Space Technology*. "The world witnessed this last October as the Ares 1-X lifted off from the Kennedy Space Center in a stunning and successful test." This is quite a remarkable assessment considering that the "stunning" test that supposedly brought us closer to Mars was a suborbital hop lasting a few minutes and using an incomplete rocket with no payload. Still, they found this meager flight experience sufficient to declare the Ares 1 "the most reliable, affordable and safest means to lift our astronauts to LEO [low Earth orbit] and beyond." Despite their advocacy for dramatic cuts in federal spending more generally, their conclusion was that "No better investment can be made than the Ares system."

Politicians were not the only high-profile figures who reacted strongly. An open letter to President Obama, bearing 27 signatures of Apollo-era astronauts and former NASA officials, lamented "canceling NASA's human space operations, after 50 years of unparalleled achievement." The letter's primary message is expressed in the following paragraph:

> We are very concerned about America ceding its hard-earned global leadership in space technology to other nations. We are stunned that, in a time of economic crisis, this move will force as many as 30,000 irreplaceable engineers and managers out of the space industry. We see our human exploration program, one of the most inspirational tools to promote science, technology, engineering and math to our young people, being reduced to mediocrity. NASA's human space program has inspired awe and wonder in all ages by pursuing the American tradition of exploring the unknown. We strongly urge you to drop this misguided proposal that forces NASA out of human space operations for the foreseeable future.

In a similar statement, Apollo astronauts Jim Lovell, Gene Cernan (both of whom also signed the above-mentioned letter), and Neil Armstrong registered their disapproval of the program changes, stating that

> ... the accompanying decision to cancel the Constellation program, its Ares 1 and Ares V rockets, and the Orion spacecraft, is devastating . . . It appears that we will have wasted our current $10-plus billion investment in Constellation . . .

The availability of a commercial transport to orbit as envisioned in the President's proposal cannot be predicted with any certainty, but is likely to take substantially longer and be more expensive than we would hope.

Across the various components of the space community, there were some common themes to the complaints from opponents of the new plan:

- U.S. human spaceflight would be cancelled, yielding America's leadership to Russia and emerging space powers like China and India. (Notably, Canada, Japan, and Europe were rarely mentioned despite their sophisticated space capabilities and their status as ISS partners.) We would be dependent on Russia for access to our own space station.
- Thousands of skilled workers at government and contractor sites around the country would lose their jobs.
- The Constellation program's Ares 1 and Ares 5 launch vehicles are the safest and most cost-effective approach to space access, and we would be throwing away a huge investment (usually quoted in the range of $9-10 billion through fiscal year 2010) if we stopped developing them.
- The U.S. private sector cannot be relied upon to provide safe, affordable, efficient launch services to NASA for its low Earth orbit needs.

Each of these points deserves closer examination because there's more to the story than the sound bites and spin are intended to communicate.

The end of U.S. human spaceflight? Obviously, this was never the administration's intent. Within four months of taking office, the president nominated Charlie Bolden, a former shuttle astronaut, to be the new NASA Administrator, and formed the high-visibility Augustine Committee to study how the U.S. could improve its human spaceflight efforts. Obama's FY11 budget proposal included extension of ISS operations beyond the time envisioned by the Bush administration, to at least 2020, performing a research program largely devoted to studying long-duration human spaceflight. Additionally, the president followed up his budget proposal by calling for human flights to an asteroid in the 2020s and to Mars in the 2030s. These are not the actions of an administration attempting to kill human spaceflight.

Shifting responsibility to the private sector for cargo and crew launches to low Earth orbit would be a boon to human spaceflight, not its death knell. For the first time, the United States, and spacefaring nations more generally, would have multiple options for putting people in orbit. This is an important milestone that should be welcomed, as it demonstrates a new level of maturity attained after a half-century of spaceflight. It also frees up NASA resources to perform research and conduct technology demonstrations rather than devoting excessive effort to operations as the agency has been compelled to do in the past.

It's possible that some people who decried "the death of U.S. human spaceflight" in their opposition to the Obama plan actually believed it to be true. If so, a big part of the reason may have been the administration's inability to adequately explain its intent or describe a fully developed vision. It's also likely that many who made this allegation were purposefully exaggerating the threat to rally the opposition, despite having a weak argument—a strategy that is not uncommon in the political realm.

Massive job loss? Sadly, yes, but as discussed in the previous chapter, that was going to happen even if the Constellation program continued, primarily in the contractor community. This is amply demonstrated by the fact that the thousands of laid-off shuttle workers were not rehired when the new heavy-lift launcher and ongoing Orion capsule programs were in full swing in 2012.

NASA didn't manufacture this problem, nor did they ignore it, but they've taken plenty of heat for it, such as Sen. Shelby's unfair accusation at an April 2010 hearing that NASA's management dismisses the job loss as "mere collateral damage."

Safety and cost-effectiveness of the Ares rockets? I will not engage in the "good rocket/bad rocket" debate, analyzing every Ares subsystem to highlight its merits or demerits and prove that it's the best or the worst rocket ever designed. That exercise was a full-contact sport in the trade press, blogs, and at conferences for long enough to qualify for inclusion in the Olympics. Not being an engineer, I will not challenge the technical judgment of the engineers who worked on Ares or those outside the program who identified problems. What I will question is the veracity of Ares proponents—many of whom were not engineers—who extolled the near-miraculous safety characteristics of a launch system that had never flown. People in the launch business know that a rocket must build a safety and reliability record over several flights, under full operational conditions.

Cost effectiveness is another area that generated more speculation than the available data warranted. It was too early to develop reliable estimates of the fixed and variable costs of operating the Ares launch system. Instead, proponents often resorted to the sunk-costs argument: "We can't waste the $9 billion we've already invested." But such an argument should not carry the day until we consider it from the flip side: how much would it cost, and how long would it take, to complete the project? If that project is defined as the Ares 1 crew launcher, the Orion crew capsule, and the Ares 5 cargo launcher, the lowest cost-to-completion estimate I heard in 2010 was $40 billion. In other words, the Ares/Orion launch system, supposedly based on proven technology and shuttle-era manufacturing capabilities, would take years longer and cost as much or more to develop (adjusting for inflation) than the original space shuttle system, which was built from scratch.

The decision-maker faces a question separate from the good rocket/bad rocket argument: Is the Ares system the right rocket, for the right mission, at the right time, for the right price? After considering the findings of the Augustine Committee and weighing other factors, the Obama administration determined that this question could not be answered in the affirmative.

Credibility of private-sector launch services? The private sector has built every rocket that the United States has ever used to launch cargo and crew to orbit or beyond, so it's safe to assume they have the necessary capability. The question is really about which particular launch operators will do the job for NASA, and how much oversight the government will have over design, manufacture, and operation of the launch vehicle.

Who will do the job will be determined by the ongoing competition between a broad range of companies from well-established aerospace behemoths to small start-ups. As in any industry, not all entrants will survive for the long haul. Some may merge or otherwise change character along the way. The situation will evolve into something that may look very different in a few years. It may or may not include the same companies that today hold agreements with NASA. But regardless of who the launch providers are, NASA will not procure their services unless the agency is satisfied that safety and reliability requirements have been fulfilled.

Positive reactions

There were many positive responses to the plan, and not just from companies holding commercial launch contracts with NASA. Not surprisingly, an industry association called the Commercial Spaceflight Federation was happy with the proposal. Linking it to important milestones in aviation, the Federation stated that "With President Obama's historic decision, we stand on the threshold of a new era in space. The commercial spaceflight industry is working to extend the legacy of the Wright Brothers into space . . . This initiative is on par with the government Airmail Act that spurred the growth of early aviation and led to today's passenger airline industry."

Many in the space advocacy community reacted favorably as well. Bill Nye, director of the Planetary Society, expressed the happiness of his organization, which is a large international group of space science enthusiasts: "We at the Planetary Society are excited about the new space policy because it's going to take us to someplace new rather than someplace old." He also conveyed his suspicions of the motivations of the naysayers: "The idea of the new space policy is to explore new places in space and continue the retiring of the space shuttle, which was started six years ago during the previous administration. Apparently it's people in the states directly affected by the cancellation of the space shuttle just raising a stink and promoting misinformation."

Although a number of Apollo-era astronauts signed letters objecting to Obama's plan, not all astronauts felt the same way. In July 2010, 22 shuttle astronauts and two from Apollo wrote to key senators on NASA's authorization and appropriations committees in support of the commercial human spaceflight plan:

> Let us be clear: we believe that the private sector, working in partnership with NASA, can safely develop and operate crewed space vehicles to low Earth orbit . . . The success of NASA's proposed Commercial Crew program is critical. By allowing the private sector to take on the transportation of crew to low Earth orbit, NASA will finally be able to direct its resources and focus on human exploration beyond, and we strongly feel this direction for the agency is the right one.

Concurrently, those same senators received a letter from five of the 13 members of the Columbia Accident Investigation Board (CAIB), the group that investigated and reported on the destruction of space shuttle *Columbia* in 2003. Many opponents had been pointing to the CAIB report as evidence that the new plan was taking NASA in the wrong direction. But these CAIB members saw it differently:

> We have also noted with interest recent space policy discussions where our report has been cited. In particular, we have been somewhat surprised to learn that some people, both within and outside of the Congress, have interpreted the new White House strategy for space which gives a greater role to the commercial sector in providing crew transportation services to the International Space Station, as being not in line with the findings and recommendations of the CAIB report. Our view is that NASA's new direction can be a) just as safe, if not more safe, than government-controlled alternatives, b) will achieve higher safety than that of the Space Shuttle, and c) is directly in line with the recommendations of the CAIB ... it has been suggested by some that only a NASA-led effort can provide the safety assurance required to commit to launching government astronauts into space. We must note that much of the CAIB report was an indictment of NASA's safety culture, not a defense of its uniqueness.

As mentioned in the Introduction, my own reaction to the proposal was positive. I was relieved that the president had spotted the opportunity and shown the fortitude to initiate the much-needed changes. The space agency, and U.S. policy-makers in general, should be celebrating a new era in which launch operations are handed off to the private sector, just as communications satellites were handed off in the 1960s. As mentioned earlier, this is an important milestone. The nation has reached the stage at which human spaceflight is transitioning from a special government-only activity to a service available from multiple providers for a variety of public and commercial purposes. When you think about the steps required to extend human endeavor out into the solar system, you realize that this development is every bit as profound as reaching the Moon in the Apollo program (though far less dramatic). To those who grew up with Apollo, that may sound strange, even profane. But we're talking about moving space activities from special, isolated, barely sustainable actions to purposeful, mainstream activities that create value. That involves taking a step-by-step approach in which every step is essential but not necessarily exciting.

John Logsdon, professor emeritus at George Washington University and one of the signers of the above-mentioned CAIB letter, said it well in a July 2010 *Space News* commentary:

> It is not surprising that those with positive memories of Apollo and with vested interests in continuing the space status quo have been so strong in their opposition to the new approach; they are defending a space effort that to date has served them well ... Trying to change that culture and thereby close out the half century of Apollo-style human spaceflight seems to me the essence of the new space strategy ... There is no commitment to a specific destination on a specific schedule; that avoids the narrowing effect that was a characteristic of Apollo. To me this is a quite sensible and easily understandable strategy, if the United States wants to be in the vanguard of 21st century space exploration . . . To me, the greatest threat to U.S. space leadership would come from our political system insisting on staying with the Apollo-era approach to the future, not from adopting this new strategy.

Muddled message

Actually, the new NASA plan introduced in February 2010 shouldn't have been a surprise to anyone who was paying attention to the situation during the previous year or two. Even before taking office, Obama had signaled his dissatisfaction with the aim of the previous administration's program, and his concern regarding its affordability and sustainability. His appointment of the Augustine Committee shortly after taking office, and his intent to use the Committee's report to influence the next federal budget submission, should have been a clue that change was in the wind. If there was still any doubt, it should have been dispelled by the release of the Committee's report in October 2009, which confirmed that NASA's exploration program, which had been underfunded since it began five years earlier, was not executable in the budgetary environment expected in the decade ahead.

With completion of the space station and termination of the shuttle program on the immediate horizon, no crystal ball was needed to forecast that a turning point was imminent. The new administration was not convinced that the right decisions had been made in the years leading up to that turning point. They sought to change that by embracing the emergence of new commercial launch capabilities and by reinvigorating technology research, which had been suffering through at least two decades of budgetary anemia. (As the National Research Council stated in a 2011 report, "it has been years since NASA has had a vigorous, broad-based program in advanced space technology. NASA's technology base is largely depleted, and few new, demonstrated technologies . . . are available to help NASA execute its priorities in exploration and space science.") The administration's prescription was a pause for reassessment before diving into a major development program like a new heavy-lift launcher.

In keeping with my motto "it's more important to get it right than to do it fast," I found this quite sensible. John Logsdon agreed, writing in a March 2010 commentary that "if the United States is indeed to sustain a multi-decade program of space exploration, accepting the core elements of the president's proposal and taking a few years to rebuild the foundation for that future effort justify the associated risks." Of course, there was a possible political motive for this course of action. Proceeding immediately with the launch development program would leverage existing contracts, rewarding Republican-leaning states such as Alabama, Texas, and Utah. Delaying the program, and opening it up to more competition, would create the possibility of spreading benefits to Democratic-leaning or swing states like California and Florida.

The administration apparently didn't anticipate the number of prominent voices whose motto is "it's more important to do it fast, and keep existing contracts and jobs right where they are." Unlike the preparations undertaken by the previous administration to get key members of Congress on board before unveiling its 2004 exploration policy, the Obama administration didn't float the necessary trial balloons on Capitol Hill. In other words, the problem was not that congress-members were blindsided, since they had every reason to expect that significant change was coming. The problem was that they had not been consulted, so they had no opportunity to shape what was coming, or at least assess the consequences for their constituencies. The proposed changes were far from radical, as some accused, but they did involve major modifications to government contracts and employment in many districts.

Another problem with the unveiling of the new plan was that it was simply embedded in the budget proposal for FY11. This allowed observers to draw their own conclusions from what they saw, without benefit of sufficient explanation from the administration. The documentation released by the White House and NASA emphasized that the new course of action would be a "game changer," but a question remained: What was the game?

The Bush administration's exploration policy had answered that question by proclaiming that we're going back to the Moon by 2020 and to Mars sometime after that. This was an unsatisfying answer because the policy did not identify a purpose that would justify such an undertaking—it failed to address the "why" question. But at least it was easily understandable, it specified a deadline that gave it some level of urgency, and it directed NASA to develop a new human-rated launch vehicle starting immediately.

Taking a different approach, the Obama plan included a slight increase in the total NASA budget to bump up spending on technology development, stimulate private-sector launch projects, and embark on the Augustine Committee's Flexible Path concept. And by the way, Constellation and its goal of returning to the Moon would be cancelled.

As mentioned earlier, the initial reactions to a presidential budget proposal are based on a limited set of questions: What's new? What's being cancelled? What's being increased or decreased? What does this tell us about the president's priorities? The answer to the "what's new" and "what's increased" questions seemed rather vague: technology investments, largely unspecified, exemplified by things unlikely to excite most observers, such as autonomous rendezvous and on-orbit fuel storage; more stimulus funding for the current round of commercial launch entrepreneurs; deployment of better Earth science sensors; and the ill-defined "flexible path" thing, which we were told was bold, new, and transformational.

The answer to the "what's cancelled" question, on the on the other hand, was crystal clear: the Constellation program, which included the Ares 1 launch vehicle and Orion manned capsule that were already under contract, and the Ares 5 heavy lift vehicle and the Altair lunar lander that would be big hardware projects in the next decade. The cancellation dominated the story in the media and in reactions throughout the interested community. Exhortations about the bold, new, transformational path failed to gain traction. The message about the president's priorities, despite life extension for the ISS and a proposed *increase* to NASA's budget, was interpreted to be a desire to scale back or even phase out human spaceflight.

By failing to adequately articulate its message, the administration lost control of it. Admittedly, selling the message would have been difficult in any case. They needed to explain why it was worth it to cause a major disruption in a five-year-old program that already had billions invested in it. Space supporters of all types were being asked to recognize the superiority of a capabilities-driven strategy over the destination-driven strategy that had been dominant for more than half a century and had gotten us to the Moon. It was necessary to spell out why this policy and programmatic shift was in the best interest of the nation's long-term space efforts. Enlightenment of this scope and depth cannot be achieved simply by releasing a budget proposal.

Realizing that more needed to be done to explain the plan and counter his critics—and to make the Florida congressional delegation happy—President Obama responded with a speech at the Kennedy Space Center in Florida on April 15, just two and a half months after the debut of his budget. He reiterated his support for NASA and for human spaceflight, and his belief that the commercial launch community was up to the task of serving NASA's needs, including manned launches. He also introduced two new items. First, the Orion crew capsule would be built as an emergency return vehicle for the ISS rather than be terminated along with the rest of Constellation. This was an initial concession to those who wanted to see existing contracts continued, and who felt that the U.S. government needed to have its own backup to commercial crew services. Second, the president returned to Apollo-era practice by declaring destinations and deadlines. Humans would visit an asteroid by 2025, achieve Mars orbit by the mid-2030s, and land on Mars shortly thereafter, at a time, he said, when "I expect to be around to see it."

These destinations and deadlines made it into the National Space Policy released by the administration two months later, at the end of June. The Moon, however, did not. In his Florida speech, the president expressed his views on a return to the Moon by remarking: "I just have to say pretty bluntly here: We've been there before ... There's a lot more of space to explore ... So I believe it's more important to ramp up our capabilities to reach—and operate at—a series of increasingly demanding targets." Overall, the president's message seemed to be riding the fence between the capabilities-driven and destination-driven strategies. At the same time, he was writing off the Moon, which is the only piece of low-hanging fruit out there, an unparalleled training ground and resource base to support future exploration that can always be found a mere quarter-million miles away. Upon hearing the speech, informed listeners could not help but feel that the president's ill-advised Moon-bashing was motivated more by a desire to undo the previous administration's exploration plan than by a thorough, informed analysis.

An asteroid visit by 2025 is also problematic. Given the space systems likely to be available by then, the president undoubtedly was setting his sights on near-Earth objects (NEOs) rather than main belt asteroids. At first, picking one or more targets would seem to be no problem, since we've cataloged more than 8,400 NEOs (as of late 2011) and over 800 of these are one kilometer across or larger. But further examination narrows down the field considerably. Mission planners who have taken an early look at the requirements believe the round trip duration shouldn't exceed about six months, so that eliminates quite a few possibilities. Also, it makes sense to pick a larger candidate because you don't want to travel that far to see something that isn't much larger than your spacecraft. And there's the inconvenient detail that the vast majority of known objects with a diameter of 150 meters or less are fast spinners, so they'd be difficult and dangerous to approach. Even the larger ones have weak and irregular gravity fields, so "landing" on one would be more like rendezvous and grappling, and astronauts on the surface would find mobility even more challenging than on the Moon. Of course, such a program would need to begin with robotic precursor missions to characterize and select possible targets, plus a space-based tracking system to identify more NEOs and more precisely define the orbits of objects already in the catalog. If 2025 is the deadline, there isn't much time to prepare.

A temporary compromise

As might be expected, Obama's harshest critics were still not mollified by his Florida speech or his new National Space Policy. Congress-members wanted to reinstate more of the Constellation program. *Space News* editorialized that "as of now the proposed effort lacks both focus and urgency." Programmatically, NASA couldn't make adjustments to existing Constellation efforts because Congress needed to approve the president's changes in a new authorization bill. Until that was done, NASA was operating under a law that forbade cancellation of Constellation elements or major reprogramming of their funding without congressional approval.

The new authorization bill that was crafted in the summer and early fall of 2010 was mostly the work of the Senate committee overseeing NASA. It was designed to strike a compromise between the president's proposal and the preferred path of the Congress. The end result was a three-year NASA authorization bill (covering FY11-13) that the president signed in October, which included an extension of ISS to 2020 and termination of the Ares

1 launcher and the lunar landing plans. For their part, the Congress got continuation of the Orion crew capsule—with its deep space capabilities intact—and immediate start on a heavy-lift booster. Unfortunately for the president's plan, the funding for the heavy booster program had to come from somewhere, and his intended increases for commercial launch programs and Earth science were his opposition's principal targets. But governing in a democracy is all about compromise, and the space agency now had permission to move ahead with its exploration program. Permission, but not funding. There was still no appropriations bill to provide NASA with the money it needed, even though FY11 had started on October 1, 2010. In the new year, this turned out to be more of a problem than expected.

A few days after President Obama had presented his budget proposal in February, NASA Administrator Charlie Bolden explained the new plan's rationale by saying that ". . . incremental changes or tinkering on the margins will not be sufficient to address current and future needs. Rather, a fundamental re-baselining of our nation's exploration efforts is needed. We must invest in fundamentally new innovations for space technology." It was tough love—he was telling a diverse community of political, bureaucratic, technological, and business interests who had been pursuing a particular track for five years that it was time to rethink everything. But it was absolutely necessary.

Although "resistant to change" is one of the dictionary definitions of "conservative," that characteristic is not limited to political conservatives. We can all sympathize with the decision-makers, agency managers, acquisition specialists, technical experts, and others who had spent years struggling down a certain path toward particular goals only to have the whole thing upended, seemingly overnight. But resistance to change doesn't just manifest itself when change arrives. There is also a tendency toward resistance to the *idea* of change—denying that change is necessary or inevitable.

In the story so far, we've seen members of both political parties labeling the president's proposed changes as shocking and radical. We've seen parochialism overcoming political ideology, with conservatives going against their strong preference for private-sector solutions in favor of government-funded jobs for their constituents. We've seen opponents of change invoking foreign threats, declaring that developments in China (or India or Russia) dictate that we stay on the traditional path lest we be overshadowed. Resistance to change undoubtedly was among the motivations behind these actions. However, in today's world it's a poor tool in the search for solutions.

Chapter 5

Goodbye space shuttle, hello . . . what?

> *America is a nation that conceives many odd inventions for getting somewhere but it can think of nothing to do once it gets there.*
> —Will Rogers, early 20th century actor, writer, and political satirist

The year 2010 ended with a compromise between the president and Congress on the future of the nation's civil space program, embodied in the NASA Authorization Act of 2010, which covered fiscal years 2011-2013. The politics of 2011 were devoted to ensuring that the authorization law was carried out to the satisfaction of those who wrote it. To help understand the stakes involved, let's examine in a little more detail what was in the law relevant to space launch and human spaceflight programs.

NASA was directed to develop a Space Launch System (SLS) capable of taking large payloads beyond low Earth orbit. In an unusual move, the congressional language (which originated in the Senate Commerce Committee) got specific about technical and programmatic details. The legislation required that at a minimum, the launcher must have:

- initial capability, without an upper stage, of lifting payloads weighing between 70 tons and 100 tons into low Earth orbit in preparation for missions beyond low Earth orbit;
- the ability to carry an integrated Earth departure stage bringing the total lift capability to 130 tons or more;

- the ability to lift the multi-purpose crew vehicle (MPCV—the Orion crew capsule carried over from the Constellation program);
- the capability to serve as a backup system for supporting ISS cargo or crew delivery requirements not otherwise met by commercial or partner-supplied vehicles.

Artist's concept of the Space Launch System (SLS). This initial configuration is designed to lift 70 metric tons to LEO. A larger follow-on version is planned to carry 130 metric tons.
Source: NASA

In the process, NASA must "ensure critical skills and capabilities are retained, modified, and developed, as appropriate, in areas related to solid and liquid engines, large diameter fuel tanks, rocket propulsion, and other ground test capabilities." In other words, the workforce and infrastructure that had been built for the space shuttle program and other current launch vehicles must be preserved. The development program must be initiated "as soon as practicable" with a deadline for operational capability of the core elements set at no later than December 31, 2016.

The intent to continue the key elements of the Constellation program is obvious, made even more so by the paragraph titled "Modification of Current Contracts":

> In order to limit NASA's termination liability costs and support critical capabilities, the Administrator shall, to the extent practicable, extend or modify existing vehicle development and associated contracts necessary to meet the requirements . . . including contracts for ground testing of solid rocket motors, if necessary, to ensure their availability for development of the Space Launch System.

Similarly, the multi-purpose crew vehicle was a born-again Orion capsule that "shall continue to advance development of the human safety features, designs, and systems in the Orion project." The requirements made it clear that it should retain capabilities planned under Constellation and not merely serve as a crew lifeboat for the ISS, as President Obama had suggested. At a minimum, it should have the capability:

- to serve as the primary crew vehicle for missions beyond low Earth orbit;
- to conduct regular in-space operations, such as rendezvous, docking, and extravehicular activities, in conjunction with payloads delivered by the SLS or other vehicles, in preparation for missions beyond low Earth orbit or servicing of assets in cislunar space;
- to provide an alternative means of delivery of crew and cargo to the ISS, in the event other vehicles, whether commercial vehicles or partner-supplied vehicles, are unable to perform that function;
- for efficient and timely evolution, including the incorporation of new technologies, competition of sub-elements, and commercial operations.

Artist's rendering of the Orion capsule at the International Space Station. If commercial development goes well, this rendezvous won't be necessary. Source: NASA

The statutory language did not make any effort to pretend that this was anything but a continuation of the Orion project, directing NASA to "continue the development of a multi-purpose crew vehicle." The capsule shares the end-of-2016 deadline with the SLS, and "may undertake a test of the transportation vehicle at the ISS before that date."

Off to a rough start in the new year

The 2010 authorization legislation required NASA to present the reference design of the SLS and crew capsule to its congressional committees within 90 days. On the appointed date, January 10, 2011, the agency submitted an interim report. Often, agencies responding to congressionally-directed actions are not able to complete their work by the statutory deadlines, so they send interim reports to the relevant committees that essentially say "we're not finished yet, but we're working on it, and here are the preliminary results." That's usually enough to satisfy the committees that their concerns are being addressed in a timely manner. But not this time.

As instructed, NASA investigated the use of existing contracts to satisfy the performance and schedule parameters specified in the law. Administrator Bolden's guidance to his teams was that system designs should be "affordable, sustainable, and realistic" and that this should apply to the full life cycle—ongoing operations as well as development and procurement. Their task was complicated by the fact that budgets through the 2016 deadline could only be guessed at. Even the budget for 2011 was still unknown despite having reached the fourth month of the fiscal year.

The 22-page interim report stated that "neither Reference Vehicle Design [for the SLS and the crew capsule] currently fits the projected budget profiles nor schedule goals outlined" and that "a 2016 first flight of the SLS does not appear to be possible within projected FY 2011 and out-year funding levels." The agency was "exploring more innovative procurement and development strategies to determine whether we can come closer to the December 31, 2016 goal."

The Senate Commerce Committee was not satisfied with this response or NASA's promise to deliver the final report later in the year. In a short statement issued two days later, the committee expressed its displeasure by reminding NASA that "the production of a heavy-lift rocket and capsule is not optional. It's the law. NASA must use its decades of space know-how and billions of dollars in previous investments to come up with a concept that works. We believe it can be done affordably and efficiently—and, it must be a priority." As if that were not clear enough, the next day Senators Bill Nelson and Kay Bailey Hutchison, members of the Commerce Committee, issued their own statement emphasizing that "The law directs NASA to build on past investments in human spaceflight by leveraging existing knowledge and assets from the Space Shuttle and Constellation Programs."

The dueling word processors of Capitol Hill

The pressure to reconfigure NASA's programs would continue throughout the year, and not just in the Senate. In the House, amid heated discussions over deficit reduction and whether to raise the national debt ceiling, six Republican representatives from states with a stake in the human spaceflight program (Alabama, Florida, Texas, and Utah) suggested in a February 7 letter to the chairmen of the House Appropriations and Commerce Committees that NASA's climate research funding should be transferred to the human spaceflight account. They lamented the "excessive growth of climate change research" which falls "outside the scope of NASA's primary mission." Apparently, none of these representatives were familiar with the language in NASA's charter, the National Aeronautics and Space Act. It lists nine objectives for the agency, the first of which is "The expansion of human knowledge of the Earth and of phenomena in the atmosphere and space." None of the objectives specifically require human spaceflight.

A few weeks later, two of the signers of that letter, Sandy Adams of Florida and Pete Olson of Texas, sent a letter to House Budget Committee chairman Paul Ryan telling him, "we believe that NASA's budget can be reduced," which seems surprising coming from representatives of two states so deeply involved in human spaceflight programs. But they were simply repackaging the suggestion made to the Appropriations and Commerce committees previously: "we believe there is an opportunity to cut funding within the Earth Science account where an overabundance of climate change research is being conducted." That "overabundance" would be cut from the budgets of NASA's Goddard Spaceflight Center in Maryland and the Jet Propulsion Laboratory in California, both of which suffer from the obvious flaw of not being located in Florida or Texas.

Adams and Olson concluded that President Obama's budget proposal willingly yields leadership to China, Russia, and India, "countries that understand the importance of human space exploration." (As others have done, they seem to have forgotten about Canada, Japan, and the member countries of the European Space Agency, the United States' partners in the space station.) "We cannot continue to accept this administration's assault on American exceptionalism and world leadership." Evidently their concern was for leadership in human spaceflight only, and not in Earth science—the importance of which is also understood by all of the countries mentioned.

Back in the Senate, the Commerce Committee was getting increasingly impatient with NASA's inability to produce final reference designs and commence work on the SLS, stemming the job loss that was already occurring in the space workforce. Senators Jay Rockefeller, Kay Bailey Hutchison, Bill Nelson, and John Boozman wrote to the NASA administrator on May 18:

> It has now been more than seven months since the NASA Authorization Act of 2010 ("the 2010 Act"—P.L. 111-267) was signed into law, and more than a month since the Fiscal Year 2011 Continuing Appropriations Act (P.L. 112-10) removed any remaining statutory obstacles to its full implementation. To this point, the National Aeronautics and Space Administration (NASA) has not made sufficient progress in carrying out the changes Congress required in the 2010 Act. Furthermore, NASA has not adequately complied with a number of reporting requirements designed to keep Congress apprised of NASA's progress in implementing the Act... As a result, we are requesting bi-monthly briefings and detailed information documenting what steps NASA is taking to comply with the law.

"More than seven months" since passage of the authorization act sounds like plenty of time to respond. But authorization doesn't pay the bills—that takes appropriations, and NASA (along with other agencies) had to wait until more than six months into the fiscal year before Congress completed that action. In the meantime, NASA had to stay at FY10 spending levels and was not permitted to cancel Constellation contracts or initiate new projects. The May 18 letter also remarks that it had been "more than a month" since the FY11 appropriations act had "removed any remaining statutory obstacles to its full implementation." Unfortunately, there are other obstacles besides statutory ones. Anyone who works in the federal budget process can verify that funding does not automatically appear in agency program accounts the day after the president signs an appropriations bill. Launching major new programs and issuing contracts generally doesn't happen within a month unless there's a national emergency. It even takes time to shut down old programs. The Constellation program wasn't officially terminated until June 10, two months after the appropriations act was signed.

There was more action in the House as the Washington summer heated up. In July, the Appropriations Committee produced a NASA bill for FY12 that would cut $1.6 billion from what the agency received in FY11. The House plan would cancel the James Webb Space Telescope, which was suffering cost overruns and schedule delays. It would also cut back substantially on the president's request for commercial spaceflight, technology development, and Earth science while increasing the human spaceflight accounts. According to the committee's report, the commercial funding request was cut because "The sizable increase in the budget request . . . was premature given the still-undefined acquisition strategy . . . and the uncertainty behind assumptions about pricing, schedule, market demand, flight opportunities and other economic factors that are essentially unknowable at this time." It's ironic that the committee pointed out these uncertainty factors as the reason for cutting commercial development, while similar factors were just as uncertain—or even more so—for the SLS and crew capsule, which they awarded large increases.

The House committee report justified a $100 million reduction to the president's request for Earth science as follows: "While the Committee supports Earth Science functions, this area has rapidly grown over the past few fiscal years, and the current constrained fiscal environment simply cannot sustain the spending patterns envisioned by NASA in this field." Once again, leadership in Earth science was not part of legislators' calculations. NASA's funding in this field shrank during the Bush administration even as data flowed in from the satellites developed in the 1990s. But those satellites had reached, or were about to reach, the end of their expected service lives and needed to be replaced, prompting the Obama administration to reinvest in this activity to maintain U.S. leadership.

The House committee report also demonstrated the power of the Congress to be as specific as it wishes in directing expenditures in agency programs—and in the use of this power by influential members to take part in regional rivalries. The report mandated that funds appropriated for the SLS and the crew capsule were intended for design and development only, not for other related expenses such as "civil service oversight, program integration, ground operations, and mission operations" which would have to find resources elsewhere or be deferred. This was a stealthy form of earmarking. The committee wanted the money to flow to the Johnson Space Center in Texas, the Marshall Spaceflight Center in Alabama, the Stennis Space Center (which operates rocket engine test stands) in Mississippi, and their contractors—not used for administrative purposes at

NASA Headquarters in Washington or for necessary facilities updates to accommodate the new launcher at the Kennedy Space Center in Florida.

Key senators weighed in on this as well. "There is an immediate need to begin design and construction efforts related to SLS, including those associated with rocket propulsion testing," Sen. Shelby wrote to the president in an August 15 letter cosigned by Republican Senators Jeff Sessions of Alabama, Thad Cochran and Roger Wicker of Mississippi, and David Vitter of Louisiana. The letter complained that SLS funding was being used for infrastructure and "general support" at the Kennedy Space Center, and charged that "efforts to spend SLS funds on priorities other than SLS violate the authorization act . . . and suggest disregard for Congress' constitutional authority." In addition to cutting Florida out of the funding stream, such language seems excessive and driven by partisanship rather than legitimate concern, as does the senators' accusation that "this administration has no intention of properly using appropriated funds."

The two senators from the Sunshine State were not about to let this assault go unchallenged. Democrat Bill Nelson and Republican Marco Rubio together sent a letter to the president on August 26, taking a more gentlemanly tone than their Senate colleagues had. They observed that "there appears to be a misunderstanding" which prompted them to write "to clarify the intent of the law." They stressed the importance of infrastructure projects such as removal of shuttle launch support structures and upgrades to the crawler transporter, characterized them as cost-saving measures, and urged the president to continue applying SLS funds to these tasks.

Another parochial battle was underway in the Senate. In this case, the issue was the SLS propulsion contract. A repackaging of the Constellation contract would have ATK of Utah producing the strap-on solid-fuel boosters for the new heavy lifter, just as the company had done throughout the space shuttle program. But some senators objected to a sole-source award to ATK, saying that the technology could be advanced and costs could be trimmed if it were open to competition. The prominent voices making this argument to the NASA administrator in May and June were California Democrats Barbara Boxer and Diane Feinstein, as well as an unusual ally, Sen. Shelby. Re-competing the booster contract would open the door for Aerojet (a California company) and Teledyne Brown (a Huntsville, Alabama, firm with its parent company in California), which had teamed up to pursue this work. Redesigning the boosters also would provide additional work for NASA design and development teams at the Marshall Space Flight Center in Alabama.

This threat to ATK's dominance sparked a quick response. Utah Senator Orrin Hatch joined with his Republican colleagues Mike Crapo and James Risch (Idaho) and Dean Heller (Nevada), as well as Democratic Senate Majority Leader Harry Reid (also of Nevada) in an August letter to NASA Administrator Bolden and OMB Director Jacob Lew calling for fast action on the SLS project, but objecting to competition on the propulsion contract. "We strongly believe conducting a competition earlier in the development of the SLS will only create further delays and cost overruns . . . it is irresponsible to spend funds on the development of a new system, such as an enhanced liquid engine, to accomplish what is already possible through existing technology, specifically solid rocket motors."

With legislators specifying the design, micromanaging the program funds, and choosing the propulsion system, it's no wonder that critics of the SLS began referring to it as "the congressional rocket" and "the Senate Launch System."

The plot thickened in late August when a Russian Soyuz rocket carrying supplies to the space station suffered an engine failure and didn't make it to orbit. It wasn't immediately clear whether the launch delays resulting from the investigation and remedial actions would create a significant snag in station operations. (As it turned out, they didn't.) Nonetheless, the same day that this occurred, members of Congress emphasized how this event reinforced their positions—including positions that were in direct opposition. Senators Hutchison and Shelby issued a joint statement on the Soyuz malfunction:

> This failure underscores the importance of successful development of our own National capabilities and at the same time demonstrates the risks with having limited options for ISS supply and crew rotation. As we have already seen with the multi-year delay with commercial providers of cargo to the space station, the country would greatly benefit from the timely implementation of the NASA Authorization Act of 2010 and development of the Space Launch System (SLS) as a back-up system . . . We strongly encourage NASA to immediately announce this week—not next month—the design for their next launch vehicle, which will halt the further loss of skilled aerospace workers now poised to be laid off from the NASA manned spaceflight program.

Since the SLS was not being designed to service the space station, and would be too large and expensive for that task, it's evident that the senators were simply using the Russian launch mishap as an opportunity to keep the pressure on NASA to move ahead on SLS. But fellow Republican Dana Rohrabacher, a House member from California with more than two decades of experience on space issues, saw things differently:

> . . . this episode underscores America's need for reliable launch systems of its own to carry cargo and crew into space. The only way to achieve this goal is to place more emphasis on commercial cargo and crew systems currently being developed by American companies.
> We need to get on with the task of building affordable launch systems to meet our nation's needs for access to low Earth orbit, instead of promoting grandiose concepts which keep us vulnerable in the short and medium terms. The most responsible course of action for the United States is to dramatically accelerate the commercial crew systems already under development.
> I am calling on General Bolden, the NASA Administrator, to propose an emergency transfer of funding from unobligated balances in other programs, including the Space Launch System, to NASA's commercial crew initiative . . . NASA could potentially transfer several hundred million dollars from this long term development concept, since the SLS project has not even started, to the more urgently needed systems that can launch astronauts to ISS, reliably and affordably.

Here we see an example of contrasting approaches to the issue from within the same party. Rohrabacher's endorsement of the commercial solution and his disdain for government-funded "grandiose concepts" (an obvious reference to SLS) is true to his conservative ideology. He reiterated his stance two months later in a *Space News* commentary in which he said, "we are wasting our borrowed money on monster rocket designs that incorporate narrow parochial political factors with a stubborn refusal to learn the lessons of the past." This put him in direct conflict with the interests of his Senate colleagues, who were working to increase the funding and pace of the booster program and would strenuously resist the shift in priority and dollars that Rohrabacher suggested.

At the time, NASA's earliest projection for the first flight of SLS was late 2017, with the first launch of a crew at least a year later. Given the high-priority needs of the ISS and the pace of commercial launcher development, Rohrabacher was proposing the right solution. In commenting on this situation, the editorial page of *Space News* agreed: "Absent a major change in U.S. human spaceflight policy, the Commercial Crew Development program represents the nation's best near-term bet for fielding an astronaut-launching capability."

Another controversial event was the September 7 appearance in *The Wall Street Journal* of leaked cost estimates for the exploration program elements. An internal NASA study outlined five options for the SLS, crew capsule, and associated ground components from 2012 through 2025. The costs differed widely due to variations in the pacing, and in some cases the content, of the development options, ranging from a low of about $41.5 billion to a high of over $62 billion. The pace specified by the Congress in NASA's authorization act was priced at $57 billion.

Senators Hutchison and Nelson fired back with a joint press release the next day, accusing the administration of engaging in a "campaign to undermine America's manned space program." In their view, NASA had invented "wildly inflated" cost numbers based on "an imaginary acceleration" which they called "a convenient myth" that no one had proposed. Sen. Shelby chimed in as well, telling *Space News* that the Obama administration had concocted the internal study and the leak to the press "in an attempt to justify its dithering and flagrant defiance of the law."

Based on the vehemence of the senators' reactions, one would think that conducting an internal study of alternative schedule and cost estimates is both unprecedented and nefarious. In fact, it's often a necessary exercise, especially at agencies like NASA and the Defense Department that carry out big, complex, lengthy development programs that must fulfill requirements in a timely manner. Sometimes, congressional committees specifically ask for such estimates. The leaked study clearly shows that NASA reviewed the administration's budget projections for the coming years and determined that it would fall short of the requirements in the 2010 authorization act. Based on those projections, the first SLS flight with a crew would come in 2021, then flights would be as infrequent as once every other year, and the full-size version of the SLS (130 tons of payload to orbit) wouldn't appear until after 2030. That budget projection not only missed the statutory deadlines, it also failed to include any investment in orbital infrastructure development, leaving crews with nothing to do once they finally did fly. Some rethinking was in order.

Senators Hutchison, Nelson, and Shelby didn't see it that way. All three pointed to a White House-requested study that had just been completed by the consulting firm Booz Allen Hamilton. According to the Hutchison-Nelson press release, the consultant study "found that development of the Space Launch System, Multi-Purpose Crew Vehicle and complementary ground system was feasible within authorized funding levels and timelines." However, this statement mischaracterizes the message of the study as expressed in its publicly released 11-page executive summary, which gives the reader a very different impression. (The senators undoubtedly had access to the full report, but it's highly unlikely that it completely contradicted what was in the summary.)

Booz Allen Hamilton was not asked to create its own cost estimates for the launcher, crew capsule, and ground systems programs. Rather, it did an assessment of NASA's existing cost estimates to judge their validity and utility. Its findings were cautionary, declaring that the three program estimates were "serviceable point estimates for budget planning in the near-term 3-5 year budget horizon as they represent the basis upon which future estimates can be constructed." But due to unjustified assumptions about future cost savings and the unquantifiable risks associated with technology development and cost growth, the Booz Allen Hamilton team viewed "each Program's estimate as optimistic . . . NASA cannot have full confidence in the estimates for long-term planning." For all three programs, the study team declined to score the executability of the schedule estimate. This is hardly a ringing endorsement, and is quite different from what the senators wanted everyone to believe.

Sen. Bill Nelson, NASA Administrator Charlie Bolden, and Sen. Kay Bailey Hutchison on Capitol Hill at the September 14, 2011 announcement of the SLS/Orion configuration. Source: NASA

Just a week later, the cost estimate incident seemed to be forgotten. Senators Hutchison and Nelson stood with NASA Administrator Bolden at a press conference announcing the configuration chosen for the SLS, which Bolden said would "create good-paying American jobs, ensure continued U.S. leadership in space, and inspire millions around the world."

Traditionally, the Congress has not been the initiator of major space projects. This is not surprising, because the Congress is not intended or equipped to be the incubator of federal technology and infrastructure programs. Its responsibility in this area is funding and oversight. As the old saying goes, "The president proposes and the Congress disposes." The impetus for major civil space projects comes from executive agencies (especially, but not exclusively, NASA), scientists in academia, engineers at aerospace contractors, and other sources—but not the legislative branch. If the Congress wishes to expand its traditional role, it's appropriate to ask whether this is necessary or desirable.

The previously mentioned May 18 letter from four senators to the NASA administrator stated that "The 2010 [NASA Authorization] Act lays out a carefully considered bipartisan vision of the best path forward for NASA." What motivations drove those considerations? What manner of expertise was brought to bear in crafting the "bipartisan vision"? By what standards is this judged to be the best path?

A number of statements in the authorization act, scattered throughout the document in sections labeled "sense of Congress," provide kernels of long-term strategy:

- ... form the foundation of initial capabilities for missions beyond low-Earth orbit to a variety of lunar and Lagrangian [gravitationally stable] orbital locations.
- ... provide operational experience, technology development, and the placement and assured use of in-space infrastructure and in-space servicing of existing and future assets.
- Technology development provides the potential to develop an increased ability to operate and extend human presence in space, while at the same time enhance the nation's economic development and aid in addressing challenges here on Earth. Additionally, the establishment of in-space capabilities, use of space resources, and the ability to repair and reuse systems in space can contribute to the overall goals of extending human presence in space in an international manner.

These statements are very encouraging, and constitute possibly the best-informed recommendations for the nation's strategy in space that can be found in any U.S. legislation in recent decades. They aim at the right objectives: developing new technologies, building mission capabilities and operational experience, creating space infrastructure, enabling in-space servicing, and harvesting space resources. This is precisely what NASA should be doing to advance the goal of moving steadily outward to the rest of the solar system while at the same time bringing benefits to Earth. But good words must be followed by adequate resources, sustained political support, and good implementation. For this, the authorization act is far less encouraging. It seems to dash hopes of achieving any of these objectives when it says, "It is the sense of Congress that NASA needs to re-scope, and as appropriate, down-size, to fit current and future missions and expected funding levels." In Washington-speak, "re-scope" as it's currently used *always* means downsize.

Further discouragement comes from the political rhetoric, which seems to ignore the statutory language quoted above and to misunderstand the trajectory that space development must follow to remain sustainable and productive. Endorsing a short-term tactical approach that continues old ways of doing business, we are told that America's "best path forward" in space is to build a new rocket as quickly as possible, and after it's done we'll figure out what to do with it. It seems that we haven't changed much in the decades since Will Rogers made the comment quoted at the beginning of this chapter. On the current path, the immediate priority is to provide temporary job security to selected aerospace workers in a few locations around the country—but at a high cost to long-term strategic efforts in space as resources and attention are diverted from the rest of NASA's portfolio of programs.

At an April 2010 hearing in which NASA Administrator Bolden was testifying, Sen. Shelby admonished him regarding the president's plan to rely more heavily on commercial providers: "We have made these mistakes before, Mr. Administrator. Albert Einstein said the definition of insanity is doing the same thing over and over again and expecting different results. I believe that is the case here." Ironically, the senator was using this familiar Einstein quote to support his preferred approach, which would have the U.S. government keep doing rocket procurement and launch operations just as it's always been done, and expect that this will somehow yield a different result, becoming sustainable and affordable. This is but one illustration of the difficulty in getting entrenched interests to embrace a new way of thinking.

As the space shuttle was preparing for its final flight in July 2011, Administrator Bolden reminded everyone that "The decision on the heavy lift is going to be a very expensive and very critical decision for the nation. It's got to carry us into this next era that we hope will extend longer than the 30 years of the shuttle." This is another way of saying what I've already said a couple of times about space exploration and development: It's more important to get it right than to do it fast. It also can be seen as Bolden's plea for programmatic sustainability and long-term strategic purpose.

Another year, and still unresolved

Ambrose Bierce, an American author and satirist who published a very cynical dictionary in 1906, defined politics as "Strife of interests masquerading as a contest of principles." The debate over the future of U.S. civil space efforts provides a fine example of what Bierce was talking about.

By the end of September 2011, less than 20 months after President Obama thought he was cancelling the Constellation program, its major components were back in business in a born-again space exploration plan. The Orion capsule became the Multi-Purpose Crew Vehicle (although the Orion name was retained), the Ares 5 became the SLS, and even the Ares 1 was resurrected as the Liberty Launch System, a private-sector effort by ATK aimed at lofting its first crew into space by 2015, which gained NASA support through a technical assistance agreement. The 2020 target date to return to the Moon, and the lunar lander needed for that mission, were missing from the new plan, having been replaced by a human encounter with an asteroid by 2025. Wrangling among members of Congress, and between the Congress and the White House, had produced something touted as a compromise: a new program that looked much like the old program, with a rationale that was just as weak, if not weaker. Still missing was an enduring purpose for the whole enterprise that was worth the price of admission. It still needed a justifiable focus, beyond short-term job creation.

The NASA budget for FY12 was passed by the Congress and signed by the president in mid-November—a month-and-a-half late for the start of the fiscal year, but pretty close to on-time by modern standards. Overall, the agency got $684 million less than what it received in FY11, and $924 million less than what the president had requested. SLS was granted $1.8 billion, a down payment of at most ten percent of the rocket's development cost. The Orion capsule received $1.2 billion. Orion had consumed about $5 billion by mid-2011, even before beginning substantive work on the life support system and service module. At that time, it was estimated that an additional $6 billion over five years would be needed.

The two budget line-items that were key to Obama's attempt to move NASA to a more forward-looking paradigm didn't fare well. Space technology development received a little more than half of what the president requested ($575 million rather than $1.024 billion) and commercial spaceflight got less than half the request ($406 million rather than $850 million). The commercial spaceflight shortfall prompted

NASA's independent Aerospace Safety Advisory Panel (ASAP) to state in its 2011 annual report that this budget "will not allow commercial crew transportation to the ISS by 2016. In fact, if the new funding level continues into the future, it is the ASAP's belief that the program is in jeopardy . . . If the development program is continued without adequate funding, it will increase the likelihood that safety-related testing and modifications to correct any design deficiencies would not be made."

At the end of 2011, opponents of SLS called it a rocket to nowhere, while its proponents insisted it was a rocket to anywhere. The hot argument in the space community was whether it was smarter to build a heavy-lift booster like SLS, or to use on-orbit refueling depots and smaller launch vehicles instead. Much of the discussion consisted of each faction telling the other, "My approach is faster, cheaper, and more efficient, and it's needed immediately. Your approach won't be needed until later, if ever."

NASA did see some positive developments in 2011, just not in its human spaceflight efforts. In August, the agency launched the Juno probe on a five-year trip to Jupiter; in November, the Mars Science Laboratory with its *Curiosity* rover took off for its mission on the Martian surface. The downside for space science was that long-term funding projections forced NASA to scale back its participation in the European Space Agency's ExoMars missions scheduled for 2016 and 2018. NASA previously had promised to provide the launch vehicles for both missions and a Mars rover for the latter mission. In February 2012, NASA backed out of the project completely.

ExoMars wasn't the only international project to experience cancellation of promised U.S. contributions in 2011. NASA's sister agency, the National Oceanic & Atmospheric Administration (NOAA), backed away from two collaborations with Eumetsat, Europe's meteorological satellite organization, a long-time partner. NOAA announced it would not be able to supply instruments for Eumetsat's second-generation polar weather satellite system, nor would it be able to provide the spacecraft and launch vehicle for the Jason-3 Earth science satellite that was scheduled for 2014.

Those with long memories who take a dispassionate look at 2011, the first year of a new decade, may find themselves asking, "Was the American space program poised for a replay of the 1970s?" There are many similarities: across-the-board retrenchment in program ambitions and resources; termination of a flagship human spaceflight capability; an expected gap of a few years before a new spaceflight capability could be ready; development of a new launch vehicle using a disproportionate amount of agency resources;

and inadequate strategic planning for what will be done with the new vehicle, or indeed, in exploration and development generally. As we harbor fears that other nations will surpass us in space, we look forward to a decade with no major U.S. human spaceflight events.

In many ways, the 1970s were a low point in American space activities, aside from some breakthrough missions in planetary science such as the *Viking* missions to Mars and the beginning of the grand tour of the outer solar system by the two *Voyager* probes. The thought of undergoing a latter-day version of the post-Apollo slump is a gloomy prospect, enough to send hard-core space geeks like me into despair. So let me attempt to end this discussion of the less-than-inspirational year 2011 on a positive note. Despite the negative aspects of the 1970s, the decade produced a lot of futurist thinking about what we could do in space once we got moving again. Sure, not all of the grand ideas came to pass—at least not as fast as proponents thought they would—but much of that thinking continues to enrich today's space innovators. Maybe a similar flowering of insight will appear in the current decade, especially as so many emerging nations and organizations join the space club.

CHAPTER 6

Transition, or more of the same?

Confusion is the hallmark of a transition.—Anne Grant, 18th century Scottish poet

Predicting the future is much too easy . . . You look at the people around you . . . and predict more of the same.—Ray Bradbury, science fiction author

William Gerstenmaier, NASA's Associate Administrator for Human Exploration and Operations, began 2012 by calling it a year of transition. He was referring to a programmatic shift away from shuttle operations and space station assembly to a new focus on ISS operations and preparations for deep space missions using new launch architectures. On the political side, however, the debate on the future of the space enterprise looked much as it had before, except that 2012 was an election year.

Space on the campaign trail

The rigid political calendar followed in the United States declares that a presidential election must be held every four years. Modern American experience seems to indicate that *every* year is a "presidential campaign year" because the battle to become the next occupant of the White House gets underway even before a newly elected president is inaugurated.

Election season inevitably has space advocates pondering questions such as: Which candidate will be a strong supporter of space exploration and development? Which one is most likely to articulate a bold new initiative for the American space program? Why don't the candidates pay more attention to space issues?

Space has never been a major topic in presidential campaigns. In fact, it rarely shows up at all on the radar screens of candidates and voters, as it competes for attention with higher-salience issues like the economy and overseas conflict. Public concern for space is minimal even though polls show that public fascination with spaceflight is substantial. At one time, civil space achievement held a higher place on the agenda of U.S. presidents because it was a very visible tool that proved useful against a geopolitical peer competitor. That state of affairs, and the existential threat it presented, was at its peak in the 1950s-60s and gradually diminished until it disappeared in the early 1990s. Since that time, presidents (and presidential candidates) have reconsidered how the civil space program contributes to their goals, and have found it wanting.

Space is a "constituentless" issue, which means that it doesn't encourage the formation of major support coalitions. Space advocates are widely dispersed, except for a few locations where space-related employment is high. Their organized groups (aside from big aerospace contractors) don't have a lot of resources, don't speak with one voice (in fact, they often disagree among themselves), and only represent a few tens of thousands of citizens. That may sound like a lot of people, but it's vanishingly small compared to influential interest groups like the environmental organization Sierra Club (1.4 million members), the National Rifle Association (nearly 4 million members), and the American Association of Retired Persons (AARP, 40 million members).

Even in the absence of a large, dynamic constituency, the space program could merit a president's (or candidate's) attention if it would help achieve important goals. Political scientists have identified three major goals that motivate presidents: getting reelected, making good policy, and achieving historical notoriety. The civil space program today is perceived as contributing little or nothing to the three goals because the benefits that can be derived from this activity (with the exception of jobs in a few specific locations) tend to be long-term, widely distributed, impossible to measure, and disassociated from their originators. The president may not get any credit for space benefits, at least not in time for the reelection campaign. As for making good policy, we've already seen that there's a lack of consensus on what constitutes good space policy, and in any case the next president is likely to review the program and change it. The space program would seem to have substantial appeal for the final goal, historical notoriety. But if the president fails to win enough support to initiate a space project, or a major project fails on his watch, the notoriety could be the type that the president does not seek.

In general, the political environment provides little incentive for political actors—including presidents—to invest their own resources in fighting for space projects, and the taxpayers' resources in paying for them. The exception is in those few localities where space can rise above the noise level. Especially in a very populous swing state. The prime example, of course, is Florida.

The quest for the Republican nomination was well underway in 2011, but the voters got their chance to participate starting in January 2012 when states began their primaries and caucuses. The Florida primary was held on January 31, briefly opening a window for space to become a newsworthy election issue. It seemed inevitable that it would, because high unemployment was the biggest issue nationwide, and Florida's Space Coast had recently suffered thousands of layoffs with the termination of the space shuttle program. Not only that, one of the candidates was former Speaker of the House Newt Gingrich, a long-time enthusiast for ambitious space projects.

A crowd of about 700 listened on January 25 as Gingrich called for a permanent American lunar base and what he described as "the first continuous propulsion system in space capable of getting to Mars in a remarkably short time." He spoke of robust activities in space science, tourism, and manufacturing. All of these are quite reasonable things that have been talked about for decades, and all will undoubtedly be part of humanity's movement into space. Gingrich's views on the prominent role of the commercial sector aligned well with President Obama's approach—he just wanted to do it faster. (Actually, so did Obama, but we've seen how Congress put the brakes on that plan.)

The essence of Gingrich's message was a sensible blueprint for the next steps in space development. However, some of the details he provided were clearly unachievable in the specified timeframe and went beyond what was seen as credible in the political rhetoric of the day. (Translation: He was widely ridiculed by mainstream media, social media, and of course, his opponents.) The lunar base, which apparently he intended as a U.S.-only effort, would be completed by 2020 without increasing NASA's budget.

He believed this could be accomplished with 90 percent private-sector investment. NASA's role would be to devote 10 percent of its budget to prize competitions aimed at program goals. (So far, NASA's prize competitions have been effective only for much smaller space endeavors, like designing a better spacesuit glove.) In the process, spaceflight would become so common that there would be "six or seven launches a day." (For comparison, the U.S. conducted 18 launches in 2011.) Eventually, when the moon base would grow to a settlement of at least 13,000 people, it would be allowed to apply for statehood. (Presumably, Gingrich expects the U.S. to withdraw from the Outer Space Treaty, which prohibits nations from declaring sovereignty on the Moon.)

Gingrich closed his speech by declaring it a historic moment, "the beginning of the second great launch of the adventure that John F. Kennedy started" at a joint session of Congress on May 25, 1961. The venue for this momentous occasion was the Holiday Inn Express off Interstate 95 in Cocoa, Florida.

In a press conference following the speech, Gingrich expressed his intention to reexamine NASA's planning, including consideration of alternatives to the Orion capsule and SLS booster. In keeping with what has become tradition, a Gingrich administration would reach for the reset button on his predecessor's civil space strategy.

Mitt Romney, the former Massachusetts governor who was the frontrunner in the Florida Republican primary, lambasted the lunar base idea during the candidates' debate the next day. He pegged the cost at "a few hundred billion dollars" and promised he would fire anyone who suggested such a thing.

A day later, January 27, Romney spoke to about 400 people at Astrotech Corp., a Space Coast company that processes satellite payloads for launch. He derided President Obama for lacking vision, and suggested that this was why the area had lost so many space-related jobs. (Another example of spinning the story to fault Obama for policies and decisions that predated his time in office.) He exclaimed to the enthusiastic crowd, "It's time to have a mission for the space program for the United States of America!" After the cheering died down, he then told them he had no mission in mind, and gave a lengthy explanation of why it would be "the wrong thing to do" to propose one at that time.

Romney's solution: appoint a commission to find a mission. It's not clear whether he realized that he was demonstrating his own lack of vision, and that he was suggesting yet another exercise in the long and rich history of space program reviews. But he did make it clear that everything would be on the table, since the commission would include "people from the Department of Defense, from the Air Force and from other branches of service, along with scientists, astrophysicists from some of the leading institutions of the world, people from the commercial sector, the industrial sector, as well as people from NASA."

The fact that NASA comes last on this list may be telling. Romney identified the objectives for the space program as follows: research on climate and natural threats from space, technology development leading to new commercial products and health treatments, and national defense. Conspicuously absent in his objectives was any mention of space science, human spaceflight, or technology development in areas that advance space exploration and development.

The same day that Romney spoke at Astrotech, an eight-person "Romney Space Policy Advisory Group," among them four space officials from previous Republican administrations and two former astronauts, issued a statement decrying President Obama's "failure of leadership." They made the curious accusation that Obama had "dismantled the structure that was guiding both the government and commercial space sectors." Without providing explanation, this seems at odds with reality since Obama didn't dismantle any part of the legal, regulatory, bureaucratic, or policy-making structure for civil and commercial space, and in fact had issued a National Space Policy in June 2010 that differed more in tone and emphasis than in content from that of his predecessor. I asked one of the group's members about this discrepancy, and he told me (from his own perspective, not speaking for the group) that "structure" in this case meant the Constellation program and the return-to-the-Moon strategy, not the structure of NASA or the space policy-making apparatus. This unusual manner of describing a program cancellation undoubtedly was lost on the casual reader.

The advisory group went on to say that Romney's "approach to space policy will produce results instead of empty promises . . . Mitt will do more than provide our space program with an inspiring vision and mission of exploration. He will also set aggressive yet achievable goals, adhere to realistic budgets, and execute on a carefully drawn plan." The timing and content of the group's comments were coordinated with Romney's campaign, appearing on his website the day of his space speech, yet they seemed disconnected. Observers could reasonably ask: How can we expect inspiring vision, aggressive goals, and execution of a careful plan from someone who just said he had no plan and belittled the notion of a lunar base?

(Ironically, one of the members of Romney's advisory group later called for a lunar base. In a conference speech on May 22, former NASA administrator Mike Griffin said that the next step in human exploration "would be to have a permanent base on the Moon." This prompted the Obama campaign to issue a press release asking whether Romney, as promised, would fire Griffin for making the suggestion.)

There was no indication that a week of public discussions on space had any effect on the outcome of the Florida primary, which Romney won with 46 percent of the vote, compared to Gingrich's 32 percent. But for a brief period in the news cycle, it subjected spaceflight ambitions to mockery, and demonstrated the lightweight role that space has on the American political scene. Space issues faded into the background after that episode, returning only occasionally in the form of sarcastic remarks from Gingrich's fellow candidates and questions from people attending political rallies. For example, a man attending a Romney campaign event in Michigan on February 24 expressed support for human spaceflight, only to have the candidate question its affordability and convey a "been there, done that" view of returning to the Moon. Romney got a laugh from his audience when he asked, "And when you get there, would you bring back some of the stuff we left?"

After more than a half-century of activity, space is still a constituentless issue that each new president must learn from scratch, only to discover that his limited constitutional powers steer him to pursue a number of minor, but flexible, changes rather than a few large, difficult ones. It may be shocking to presidential candidates and their supporters, but there is a significant gap between what the president would like and what he can do, and that gap widens during difficult economic times.

Over the years, the space community has frequently heard a refrain something like this: "If only the president would show more leadership, the space program would achieve more and advance more quickly." This statement has been aimed at President Obama just as it has been directed at his predecessors throughout the post-Apollo era. As part of the response, the Obama campaign released a fact sheet touting the president's achievements in civil space programs, mostly aimed at Florida's Space Coast (though this didn't come until four months after the Republican primary had raised the issue in that state). But in general, negative circumstances like a sour economy, shaping events like natural and man-made disasters, and public indifference give the president incentives to distance himself from the space program, not rally to its aid. In the final analysis, presidential support for the U.S. civil space program is necessary—but it's not sufficient to ensure success. (To get a thorough history on this, read *Spaceflight and the Myth of Presidential Leadership*, which is listed in the bibliography.)

Budget time again

Depending on one's perspective, NASA either dodged a bullet or was programmed for failure in the president's FY13 budget request, delivered to Congress on February 13. The agency's top-line budget proposal of $17.7 billion took a hit of only about $59 million compared to FY12 spending, a decrease of only a fraction of a percent. Not bad, given the budget squeeze of the time, although that proposal would put NASA at its lowest funding level since 2008.

The funding recommendation for the space shuttle account, still on the books to handle the program's close-out costs, shrank by nearly a half-billion dollars to just $70 million. That allowed some other programs to request increases. One of the proposal's biggest beneficiaries was the commercial spaceflight program, which tried again to achieve a funding level ($830 million) close to the amount that had failed to pass the previous year. Also, the James Webb Space Telescope sought an increase of over $100 million to compensate for chronic overruns in the program and keep it on track for a 2018 launch.

On the downside, the proposal would cut the planetary science budget by more than 20 percent, curtailing planned Mars projects and foreclosing the ability to start new flagship missions to the outer planets. The community affected by these cuts wondered why its successes of recent years were being punished. The Jet Propulsion Laboratory (JPL) in Pasadena, California, the focal point for most of the planetary program, suddenly had concerns for its future. The response from Rep. Adam Schiff, a Democrat whose district includes JPL, was: "I oppose these ill-considered cuts and I will do everything in my power to restore the Mars budget and to ensure American leadership in space exploration." Many others expressed similar sentiments.

The budget also presented notional amounts through FY17, essentially unchanged from the FY13 numbers. The accompanying statement by NASA Administrator Charlie Bolden closed by saying that the "2013 budget implements President Obama's vision for an American space program with much greater capabilities than it has today and the flexibility and determination to reach new destinations with human and robotic explorers. Our plan sets us on a path as a nation to achieve even greater goals and to make life better around the world as we strive to meet these grand challenges." This statement aligns with prior administration policies and pronouncements that attempt to blend capabilities-driven and destination-driven strategies.

It was probably not a surprise to the administration that the battle over the SLS and Orion programs was not over. Their combined FY13 request was $2.365 billion—a hefty sum, but a cut of just over $337 million from the previous year. NASA didn't expect this cut to affect program schedules. Nonetheless, some familiar voices expressed displeasure with the cut and linked it to what they saw as an inappropriate increase to the commercial launch program. On the same day as the budget release, Sen. Hutchison issued a press release saying that "Despite repeated assurances from NASA and White House officials . . . vehicle development for the heavy lift SLS rocket and the Orion capsule is cut by hundreds of millions of dollars. These reductions will slow the development of the SLS and the Orion crew vehicle, making it impossible for them to provide backup capability for supporting the space station." The ISS backup argument was a curious one for the senator to make, despite language to that effect in NASA's 2010 authorization bill. By this time, it was well known that the first crewed flight of SLS/Orion was planned for 2021, the year after ISS was scheduled to cease operations. Even if the SLS could be ready in time—which is possible, since NASA has suggested it could fly with a crew as early as 2019—it couldn't launch frequently enough to cover all of the space station's servicing needs.

Sen. Shelby weighed in on February 22 when he told *Space News*: "The President's proposed budget for NASA underfunds the Space Launch System next year and instead gives that money to speculative 'commercial' providers that continue to over-promise and under-deliver. I expect that, once again, Congress will have to force the Administration to invest in a real exploration program that adequately funds SLS."

Budget proposals are always followed by hearings, and when the Senate Commerce, Science, and Transportation Committee met on March 7, Sen. Hutchison made the additional claim that the prospects for a manned Mars mission were undermined by the level of funding sought for the commercial crew program. "It's in the numbers and it's irrefutable," she told Administrator Bolden. "The NASA administration has drug its feet on the other human space flights that we have on our agenda." She didn't specify what these other flights were, but it would be quite a stretch to suggest that human flights to Mars were "on our agenda."

The same day, Bolden also heard criticism of the commercial crew program from the House Science, Space, and Technology Committee. For example, Rep. Eddie Bernice Johnson, a Democrat from Texas, worried about subsidizing private companies when parties other than NASA, such as foreign astronauts and space tourists, would be the beneficiaries: "I can't justify to my constituents the expenditure of their tax dollars so that the super-rich can have a joy ride." The committee's chairman, Texas Republican Ralph Hall, expressed his persistent beliefs that there would be no market for commercial crew vehicles beyond NASA-funded flights to the ISS, and that the private sector's ability to operate safely was questionable.

Bolden defended the commercial crew budget, suggesting that it would be unwise to delay the program or prematurely reduce the number of companies in the competition. "Any subsequent reductions from what the president has requested for Commercial Crew only serves to delay the amount of time that we have a commercial and American capability to get our crews to the International Space Station." Nonetheless, congressional budgeters, particularly in the House Appropriations Committee, rejected NASA's judgment and pushed for immediate selection of a single commercial crew contractor. (This position received the blessing of Apollo astronauts Neil Armstrong, Gene Cernan, and Jim Lovell in May 2012, despite their acknowledgement that they "are not experts in contemporary government contracts" and were "generally unfamiliar" with the commercial crew agreements in place at the time.)

The initial round of hearings gave way to the months-long struggle to agree on a final budget in an election year. As in the previous annual cycles we've reviewed, the process provides ample confirmation of the old adage that you can't please everyone. Arguments persisted over matters thought to be settled. Partisan lawmakers and commentators continued to try to pin the blame on President Obama for a human spaceflight downturn initiated by his predecessor. And in an era when the general trend in the Congress was toward increasing partisanship, parochialism proved it could still exert a dominant influence.

If we look beyond the environment of political stalemate that characterized 2012, we find that it actually was a year of transition for emerging commercial space efforts. Many companies made progress, and none more visibly than SpaceX, which successfully delivered cargo to the space station and returned cargo to Earth with its Dragon capsule. It was the first private venture to do so, prompting many observers to declare the May 25 docking as a historical event. Whether or not it's remembered as such, it certainly had the near-term effect of countering some of the pessimism about the ability of the private sector to deliver.

Challenging conventional wisdom

There are two long-held beliefs about the political fortunes of the U.S. civil space program. The first is that it has a long history of bipartisan support. It has even been said that the space program is immune to partisanship, or at least it was until recently. This impression arises because traditionally there has been little in the way of organized efforts by political parties to engage in widely publicized battles over civil space policy.

The second belief, a complement to the first, is that congressional votes on space have always been driven primarily by geographic interests—constituent jobs and federal contract dollars in particular locations. Clearly, there were parochial influences on the placement of NASA facilities that were built after the agency's creation in 1958. For example, that's why the Johnson Space Center, which houses mission control and astronaut training, is in Texas instead of in Florida, where those functions could have been co-located with the launch site. It's also true that many legislators over the years have been very protective of their local NASA centers and contractors.

When I did my dissertation research in the mid-1990s, I was surprised to find clear evidence that these two conventional myths are flawed. The truth about their general applicability is a bit more complicated. In regard to partisanship, while space has enjoyed support from people on both sides of the aisle, partisan voting blocs on space issues have been around all along, although their character has evolved and their visibility has increased in recent years. As for the supposed dominance of parochialism, there have been many members of Congress who have regularly voted in ways unfavorable to the NASA interests in their state, as well as members who routinely cast pro-space votes despite having no major NASA-related activity back home.

As part of my study of the post-Apollo handling of the space program by the presidency and the Congress, I performed a statistical analysis on a series of congressional votes from 1970 to 1994. The results showed that partisanship was by far the largest and most consistent factor driving members' votes on space issues. In contrast, parochialism had weak to moderate influence on voting until the late-1980s. At that point, it seems to have become more prominent as a result of two occurrences: increasing annual spending on the space station program ($900 million in FY89, rising to $1.9 billion by FY91), and the waning of the Cold War after the Berlin Wall fell, which could have prompted some members of party voting blocs to shift their allegiance to state blocs (and thus parochial issues).

The votes I chose for my study involved a series of congressional actions having major ramifications for the future of NASA's largest programs. All but one of the votes were related to amendments designed to cancel the shuttle or space station program or to transfer significant amounts of funding from NASA's appropriation to other agencies.

Ultimately, all of the votes in the sample were decided in NASA's favor, which contributes to the perception of bipartisan support. But digging into the details tells a different story. All of the amendments seeking to slash NASA programs were sponsored by Democrats. In every case, a higher percentage of Republicans gave supportive votes to NASA than was the case among Democrats. In fact, more than two-thirds of the votes were "party votes" in which a majority of Republicans voted against a majority of Democrats.

Many members—10 senators and over 100 representatives—were in office long enough to participate in all of the votes in the sample, which spanned 25 years. Among these members of long tenure, it's noteworthy that all the Republican senators and most Republican representatives voted pro-NASA almost all the time, whether or not they had NASA interests in their state or district. The same couldn't be said for the Democrats in this group, who often cast a majority of their votes in a manner unfavorable to NASA even if they represented areas with NASA field centers or space-related industry.

If members were taking their cues from partisan constituencies, this contrast shouldn't be surprising. Polls have shown that Democratic voters are more likely to question the value of the space program and Republican voters are more likely to support it. But for issues that are complex or don't have high salience for their constituents, elected officials are inclined to vote based on their ideological stance or the advice of their fellow partisans. Space issues fit this mold, since they're technically complex and usually seen as low salience.

Republicans tend to be pro-business, and the majority of NASA's funding goes out to private-sector organizations, many of which also happen to be defense contractors, another favored group among conservative legislators.

Some Democrats may see the link to the defense community as a reason to question the stated objectives of civil space programs, but in general, there doesn't seem to be an inherent dislike of space among Democrats. Rather, the party's members lean toward their own ideological and constituent interests, putting other issues (e.g., social programs, environmental protection, medical research) ahead of space in the competition for scarce resources. For most of the space age, the dilemma for pro-space Democrats has been that space spending began to grow at the same time as the Great Society programs of the 1960s, which also coincided with the beginning of continuous and growing deficit spending. Space became one more player in a zero-sum game.

There's more evidence of the long shadow of partisanship than just my own research. The stories behind the space shuttle and space station programs demonstrate strong partisan influences going back more than four decades.

Throughout the Apollo years, many members of Congress, especially Senate Democrats, resented the attention and budgetary priority that the space program received, but found there was little they could do to restrain the high-visibility Moon program that was supported by party leadership. But soon after the first Moon landing, public support for space spending began to fade and attention turned to other priorities such as the Vietnam war, civil unrest, and the economy. Space opponents took the opportunity to put up roadblocks to prevent large new space ventures from being initiated.

The Senate opponents of the space shuttle, led by Democrats Walter Mondale of Minnesota and William Proxmire of Wisconsin, found many allies, including a key player in the House, Rep. Joseph Karth, another Democrat from Minnesota and the chairman of the Space Science and Applications subcommittee. A number of attempts were made during 1970-72 to use votes on authorization and appropriations bills to kill shuttle and space station proposals. Although none of these succeeded, some were close calls, reflecting an anti-technology bias in part of the electorate aligned with Democrats at that time.

Efforts spearheaded by Democrats during this period demonstrated that lawmakers were not willing to buy into long-term commitments for a shuttle and space station, coupled with plans that would ultimately lead to human missions to Mars. The shuttle and station were approved despite this opposition, but proceeded far more slowly than their proponents had hoped. This was especially true of the space station, which held a precarious position in the Congress throughout its formative years. Senator Proxmire was still around for these debates, and in 1987 tried to terminate the space station program that President Reagan had approved in 1984. The station survived, but its budget was a political football for the next few years and the program had to withstand annual votes aimed at its demise.

Deliberations over the FY89 budget illustrate the games being played with the space station program. The House and Senate couldn't agree on a funding level, and the gap between them was the difference between continuing the program or cancelling it. The 1988 presidential election was coming up, so they "passed the buck" to the next president, a move that had clear partisan implications and was designed to deflect criticism from the Congress regardless of the outcome. It was assumed that, if elected, Republican candidate George H.W. Bush would continue the station, whereas Democratic candidate Michael Dukakis would shut it down. The House and Senate compromised at $900 million, not far below the president's original request, but there was a catch: $515 million of the total would not be released to NASA until May 15 (seven and a half months into the fiscal year) based on a decision by the new president to either continue or terminate the program. If the new president killed the program, he would take the blame. If he continued the program and it faltered during his time in office, he also would take the blame. If he continued the program and it succeeded, Congress could claim at least partial credit.

If Republicans have demonstrated greater and more consistent support for major NASA programs, does this mean that the key to aggressive development of space is Republican control of the Congress and the White House? The evidence doesn't support that conclusion. The situation is fluid; priorities change and sometimes party positions flip completely as years pass. Congressional Republicans are not as cohesive a voting bloc as they were years ago, when partisanship and internal divisions were less intense and the party expended more effort maintaining a unified minority (especially in the House). Recent trends toward international cooperation in space may not appeal to the most conservative members of Congress if they appear to conflict with concerns about foreign threats and technology transfer. Eagerness to reduce the size of government and balance the budget will not spare the civil space enterprise, even if it means denying contracts to some big businesses. On the Democratic side, the party no longer embraces the anti-technology fervor that held some sway during the late 1960s and into the 1970s. Perhaps current

Democratic thinking on the space program was best expressed by Rep. Barney Frank of Massachusetts in an April 2012 online debate hosted by the Huffington Post on whether a human mission to Mars should be initiated. Rep. Frank based his position on the opportunity costs of such a venture:

> I very much favor and have voted for funds for the scientific exploration of Mars, as part of a space program dedicated to the advancement of science . . . Obviously I have no objections in principle to sending human beings to Mars, and I recognize that there is some scientific benefit from it. But the main reason to send people to Mars—attested to by most of the scientists with whom I have spoken—is simply to show that we can do it as a matter of national achievement, and what we will learn will mostly be how to do what we are doing. That is, sending people to live on Mars is not likely to produce the benefits for those living on Earth anything like what we can do by health research, cleaning the air, providing a safer environment, etc. The overwhelming majority of scientists I have spoken to have told me they believe that space exploration will be less productive from the standpoint of improving our knowledge of the universe and yielding benefits that will have tangible impact here at home, if we direct substantial amounts to human space travel.

As we've seen in the last four chapters, strong voices for parochial interests in both parties are proving to be more potent than partisanship as we move farther away from the Cold War era and domestic issues like jobs and the economy take precedence in space policy and budgeting. And today's partisan positions on space seem to be less an expression of ideology—although that still exists in some cases—and more an effort to simply contradict the opposition. Democrats express displeasure with the civil space plans and budgets of a Republican president, but don't offer more sophisticated or more viable alternatives. The situation reverses, and Republicans object to a Democratic president's handling of the space program, but they don't have any better ideas either. Notably, some Republicans heatedly espouse positions based on parochial interests that are in direct opposition to their conservative ideology, preferring increased government spending on a

NASA launcher rather than an opposition president's plan for a managed transition of launch responsibilities to a willing private sector.

Civil space research and development, including programs that aim to create new space applications with potential economic benefits, are treated as domestic activities (despite their strong international components) and carry little clout compared to, for example, the Defense Department or the Social Security Administration, which account for vastly larger segments of the federal budget. Civil space is an esoteric undertaking that's not generally perceived as part of the mainstream. As long as this remains true, the U.S. civil space effort will feel like a pawn in a game where the bigger chess pieces routinely push it around. So it's time for space to make its way into the mainstream, where it will profoundly and visibly affect humanity's future.

Chapter 7

Planes, trains, automobiles, and spaceships

History is not, of course, a cookbook offering pretested recipes. It teaches by analogy, not by maxims.—Henry Kissinger

The previous chapter questioned conventional beliefs on the roles that partisanship and parochialism have played in decision-making for the U.S. civil space program. This chapter examines the historical record to tackle another disputed bit of wisdom: the notion that large endeavors are carried out by *either* the public sector *or* the private sector, and that the two are always competitors or adversaries. Examples from American transportation industries show that the two sectors actually work in complementary ways. While they do experience friction as they try to balance their roles, neither sector could be nearly as successful without the other.

As we attempt to craft a productive path for space development, we find ourselves embroiled in disagreements over how responsibilities should be distributed or shared between the public and private sectors, whether or not government subsidies are appropriate, and how to establish a regulatory regime that is effective without stifling the growth of industry. At one extreme, people want the government to take on full responsibility for all things space, because commercial efforts will be inadequate or unsafe. At the other extreme, we're told that the government can't do anything right, or at least can't do anything efficiently, so the government should turn over much of its space portfolio to the private sector. At either extreme, legitimate issues too often are clouded by cynical attitudes or refusal to accept evidence that contradicts a preferred worldview.

If we are to have a chance of success in space development—with its high cost, high risk, and long lead-time—we had better recalibrate our thinking to something more rational and balanced. History tells us that the public and private sectors must behave as partners. Each sector has its strengths that will determine its role in the creation of space infrastructure and the realization of a space economy. The private sector, at its best, excels at developing and marketing products and services attuned to the demands of a community of users—even a very large and diverse community—and also at minimizing costs in the interest of maximizing profits. For its part, the government sector has repeatedly proven its value in the following areas:

- Funding and/or performing basic research and development that is likely to be underfunded by the private sector.
- Building or encouraging the establishment of infrastructure, such as seaports, airports, spaceports, and tracking facilities.
- Becoming an early adopter of new products and services, helping to stimulate and stabilize the initial market.
- Enacting and enforcing regulations in areas such as public safety, worker safety, and environmental protection.

Working in concert, the public and private sectors have brought immeasurably large benefits to the nation despite their occasional failures. In the absence of their cooperative efforts, infrastructure elements like those discussed below would have experienced development cycles that were more sluggish, less ambitious, and more disjointed, resulting in dramatically slower growth of the economy.

The choices are not black and white. Neither the government nor the private sector has all the answers, but it's a difficult exercise to find the right balance. That balance is not static because needs change, markets and costs fluctuate, and the availability of government funding at all levels can vary substantially. The recent trend, as noted in Chapter 1, is toward more private-sector involvement in the ownership and sustainment of infrastructure, not just its construction. This is true at the local and regional level for things like roads, bridges, tunnels, electric power plants, and water filtration facilities. It's also true for things that are connected globally, like airports and seaports. The vast resources needed to build and maintain a space infrastructure make it likely—in fact, imperative—that the space community embrace this trend.

Embrace it, but without getting carried away. Concerns about deficit spending may prompt an eagerness to pull the plug on federal involvement and support, and hand things over to state and local governments and the private sector, before adequately considering whether sub-national governments have the requisite resources and expertise, or whether the private sector believes a profit can be made. Of course, state and local governments could be left to solve this dilemma on their own by substantially raising taxes and aggressively pursuing other revenue schemes, and private-sector participants might reduce or eliminate their expectations of profit—but don't hold your breath waiting for that to happen.

There are lessons to be learned on finding the balance from the experiences of big infrastructure activities and their associated industries. They share with spaceflight the need to conduct research and development, to rely on the government to be a key customer (at least for a while), and to ensure safety and product quality through independent oversight. Valuable lessons applicable to spaceflight become apparent when we look outside the space community at how the various modes of transportation we all take for granted became what they are today through federal involvement and cross-sector collaboration.

Aviation is often seen as the older sibling of spaceflight, while ships, trains, and automobiles seem to be no more than distant cousins. All of them have something to teach us about managing growth, technological advancement, and regulation. Collectively, they have a long and rich history of carrying people and cargo, and in the regulation of safe operations. Like spaceflight, they have depended on technological developments, experimented with a wide range of vehicle designs, gradually increased their capabilities and services, and been forced to address the risks of casualties and property damage. It turns out that the arguments we're having today are not new, nor is the notion that government and industry need to work in complementary ways if there is to be progress. Teamwork across these sectors has always been a prerequisite for success, even in the presence of occasional battles between the two. It's important to recognize that the lessons of history are almost as unforgiving as the laws of physics, and ignoring or misinterpreting them will chart a path to waste and delay.

Maritime transportation since the introduction of steamships

Government assistance, Early American style. Just as U.S. government agencies today have a policy mandate to encourage and facilitate the emergence of the commercial space industries, the young United States put a high priority on encouraging maritime industries, thereby expanding trade. The first U.S. Congress (1789-1791) enacted measures to promote American shipping and shipbuilding. These included preferential taxes and duties for U.S. ships, and projects to improve navigation channels. Over the years, these incentives grew to include direct subsidies to fishermen that helped to dramatically increase the size of the fishing industry. Further recognition of the federal government's role in facilitating maritime operations came with the General Survey Act of 1824, which gave the U.S. Army Corps of Engineers responsibility for water navigation and began a long partnership with the private sector for the establishment and maintenance of port facilities. These types of assistance are analogous to support available today to emerging space industries, such as access to government launch and tracking facilities, limited indemnification against third-party damage claims, guaranteed government purchases of data and services, and development assistance from federal agencies for commercial projects that are expected to serve those agencies.

Steam-powered ships appeared in America at the beginning of the 19th century, notably with the first commercial service initiated by Robert Fulton's *Clermont* in 1807. In the decades that followed, steamboats rapidly were adopted for carrying passengers and freight through U.S. rivers, canals, lakes, and coastal waterways. The first steamship crossing of the Atlantic occurred in 1819, going from Savannah, Georgia to Liverpool, England in 21 days.

The U.S. government's efforts at facilitating maritime infrastructure development and providing subsidies had a mixed but ultimately very positive outcome for commerce. But perhaps the greatest accomplishments of the public-private collaboration were those in the realm of safety regulation—although not until after the nation traversed a long learning curve that included far too many casualties. In a foreshadowing of many tragic events to come, the city of Philadelphia conducted the nation's first official investigation of a steamboat accident in 1817.

The long gestation of safety regulation. The 19th century Congress struggled with an issue that is still familiar today: how to ensure public safety without stifling an emerging industry that will be valuable to the economy. Despite a reluctance to impose burdens, it became evident that maritime accidents, especially those caused by boiler explosions, were an increasing problem that would not be solved by an array of inconsistent and possibly conflicting regulations enacted by states and municipalities.

Steam engines enabled ships to achieve greater speed and independence from the wind, but this came at a price, beyond simply the need to use valuable cargo space to carry fuel. Boilers that were poorly designed, constructed, or maintained, and those that were pushed too hard by their operators, had a tendency to explode. The regularity of deadly accidents caused a public outcry.

Steam engines were the rocket engines of their day, in the sense that they could concentrate unprecedented amounts of energy into a device small enough to be used for vehicle propulsion. But a variety of problems could cause operators to lose control of that energy and the vehicle could become a mobile destructive force. Keen awareness of that risk in the age of rocketry prompted the U.S. government to establish a safety regulator for commercial launch services in 1984, when the private sector was just beginning to move into the launch business. So far, commercial launch providers have experienced no third-party casualties or property damage. Steamship operators, on the other hand, did not have a spotless record, and government action to address safety problems was slow in coming.

It wasn't until the early 1830s that Congress began serious efforts in maritime safety regulation. In the first half of 1838, over 300 people died in three separate boiler explosions, providing further impetus to the passage, in July of that year, of the first legislation designed to "provide better security of the lives of passengers on board vessels propelled in whole or in part by steam." Requirements included biannual boiler inspections, annual hull inspections, employment of sufficient numbers of competent engineers, and provision of safety equipment such as fire pumps, hoses, signal lights, and lifeboats.

This law, along with several amendments that followed, proved ineffective because the government was not properly equipped to monitor and enforce its provisions. Several hundred more deaths resulted from boiler explosions before Congress acted again in 1852.

The Steamboat Act of 1852 improved oversight mechanisms by creating nine districts, each headed by a supervising inspector appointed by the president and confirmed by the Senate. But the Treasury Department, which took over responsibility from the Justice Department, had very little power over the district supervisors, and enforcement remained inconsistent. Another problem was that the law only applied to steamboats carrying passengers, not to cargo vessels or tugboats. Many more lives were lost before the end of the 1850s; for example, 175 deaths occurred in three explosions just during 1858-59.

No regulatory revisions were undertaken during the Civil War. Immediately after the war ended, the nation witnessed what remains to this day the worst maritime disaster in U.S. history. On April 27, 1865, the steamship *Sultana* exploded on the Mississippi River near Memphis. Authorized to carry 376 passengers, the *Sultana* on that day was carrying over 2,400, most of them Union soldiers who had just been released from Confederate prisons and were returning home. The death toll from the explosion, fire, and drowning was 1,800—about 300 more than the much better known *Titanic* incident that took place 47 years later.

Explosion of the steamship Sultana on April 27, 1865, resulting in 1,800 deaths, as illustrated in Harper's Weekly on May 20, 1865. This is what can happen when regulations are inadequate or poorly enforced. Source: Library of Congress, Prints & Photographs Division

It was a disaster waiting to happen. In addition to being grossly overloaded, the *Sultana*'s safety equipment consisted of just one lifeboat and 76 life preservers. An inquiry revealed that the ship's boiler had been under repair prior to its final trip, but the captain, in his eagerness to get the ship back in business, cut short the repairs and likely made the explosion inevitable. In the nearly three decades since the first steamboat safety legislation, little progress had been made due to lax enforcement.

An 1871 statute established the Steamboat Inspection Service and created the position of Supervisory Inspector General, in charge of the nine district supervisors. Still overseen by the Treasury Department, authority was broadened and, the following year, shipping commissioners were placed at key ports, establishing a regulatory environment that began to resemble what we have today.

In 1884, the Treasury Department set up the Bureau of Navigation and assigned it duties that included enforcement of rules on construction, operation, equipment, inspection, safety, and documentation. Responsibility for shipping regulation moved to the Department of Commerce and Labor when it was created in 1903. But the process of developing an effective safety regulatory regime was still far from complete.

Over 1,000 people died in a fire aboard the *General Slocum* in New York's East River on June 15, 1904. In a demonstration of the inadequacy of the inspection regime, an investigation found that the disaster could have been avoided if the vessel had followed regulations on fire suppression equipment and training. This damning finding was made worse by the fact that the *General Slocum* had passed inspection a month earlier despite its deficiencies. The result of this incident was additional requirements for fire suppression and rescue equipment, and a major shake-up in the Steamboat Inspection Service, including the firing—by President Theodore Roosevelt—of the inspectors who had passed the *General Slocum*.

The Motor Boat Act of 1910 expanded regulatory authority beyond common carrier ships to recreational and commercial motor boats over 40 feet in length. It also extended government authority to vessels with propulsion systems other than steam. This Act was amended in 1940 to expand specifications for safety equipment and assign enforcement responsibilities in cases of reckless or negligent operations.

More maritime safety legislation was introduced in the Congress following the deaths of 812 people in the July 1915 sinking of a chartered steamship in Chicago. However, nothing new of significance in this area would become law until prompted by a spate of deadly, high-visibility incidents in the mid-1930s. A series of bills in 1936-37 expanded requirements for equipment and crew training, and finally required that all casualties involving regulated vessels be reported and investigated.

During World War II, President Franklin Roosevelt used an Executive Order to transfer responsibility for maritime safety regulation to the U.S. Coast Guard, putting it under a single authority for the first time. That assignment was made permanent in 1946 and holds true today.

This short summary of the evolution of maritime safety regulation and enforcement illustrates a process that took more than a century and witnessed many thousands of casualties. Despite the carnage over an extended period of time, maritime travel in its various forms survived, and increasingly thrived as it got safer. However, the commercial spaceflight industry should not take solace in this long period of tolerance to accidents. Travel by watercraft has been around since early humans figured out that wood floats. It has provided essential service since antiquity, so it's not going away. Spaceflight, in contrast, is relatively new. It can be argued that access to space has become an essential service for critical economic and security assets—satellites for communications, navigation, and Earth observation—but such an argument can't be made yet for human spaceflight, especially space tourism. It will be critical to minimize flight mishaps and maintain a safety regime that instills confidence in the users of space systems and uninvolved third parties who may be affected.

Aspiring space travelers may compare spaceflight to modern examples like the cruise line industry. Despite a few high-profile incidents that have resulted in more inconvenience than casualties, the cruise industry asserts that they have the best safety record in the travel industry. In U.S. waters, the Coast Guard has demonstrated successful oversight of cruise ship safety even as the industry has experienced impressive growth. (The number of passengers traveling aboard the world's cruise lines was about 500,000 in 1970, increasing to 14.3 million by 2010.) The Coast Guard inspects new cruise ships before they enter service at a U.S. port, repeats inspections on a quarterly basis, and has the authority to require correction of any deficiencies before allowing a ship to take on passengers at any U.S. port.

Another lesson for the commercial spaceflight industry lies in the evolution of international maritime regulations. Enforcement of rules in territorial waters has an extensive history and has long been accepted by mariners as part of their operating environment. However, the high seas traditionally were considered an area where freedom reigned, prompting many seafarers around the world to raise objections to the imposition of rules and standards in the 20th century. Eventually, through the implementation of international regulations, it was realized that adherence to an agreed-upon body of behavioral norms on the high seas was in the best interests of everyone. Similarly, today we find general acceptance of regulation applied to the launch and reentry phases of spaceflight, but industry opposition to regulation of on-orbit activities. As on-orbit commercial activity develops, especially involving human presence, the industry will come to the same conclusion that the shipping industry did in the wake of international regulations. Commercial spacefarers must recognize that theirs is currently the only transportation industry in which the departure and return are regulated, but not the journey in between, a situation that will not last for long.

Government assistance, modern style. At the beginning of the 20th century, the U.S. found itself falling further behind other nations, particularly Britain, in its status as a maritime power and its ability to produce metal-hulled ships. The Shipping Act of 1916 was an attempt to correct this deficiency and also boosted mobilization for America's eventual entry into World War I. The result was an increase in the U.S. portion of world shipping capacity from seven percent in 1914 to 22 percent in 1920, at which time the U.S. operated the world's largest merchant fleet. The government owned most of this fleet, however, and had no desire to maintain it in peacetime. The Merchant Marine Act of 1920 (the Jones Act) provided for the privatization of the fleet. It also reaffirmed and strengthened the longstanding practice of cabotage, which reserved all coastal trade to ships that were U.S. built, owned, and crewed. (In the space age, the Jones Act is analogous to the 1984 legislation allowing the private operation of space launchers.)

Not all government attempts to stimulate the shipping industry were successful, or even well-conceived. In 1928, the Congress passed legislation providing the merchant marine with a package of indirect subsidies, the most noteworthy of which involved contracts to carry the mail. But by the mid-1930s, several investigative panels found that instead of having the intended effect, the subsidies were promoting graft and inefficiency. Other examples of less-than-successful legislation include various laws, such as the Cargo Preference Act of 1954, that gave preference to U.S.-flagged ships for certain kinds of cargoes (not unlike the long-standing practice of launching U.S. government satellites on U.S. rockets). These laws had limited impact on the strategic fortunes of the U.S. merchant marine.

Other government efforts, like the Merchant Marine Act of 1936, proved more effective. With the arrival of the Great Depression, U.S.-flagged carriage of U.S. foreign trade sagged, falling from 51 percent in 1922 to 33 percent in 1933. The Congress wanted to take steps to ensure that the nation had an adequate merchant marine, and the administration of Franklin Roosevelt concurred. The administration believed that if subsidies were necessary, the government shouldn't try to hide them or employ ineffective measures. Rather, they should be direct subsidies that are openly acknowledged.

The Merchant Marine Act of 1936 took a protectionist approach that involved both direct and indirect subsidies with the hoped-for result of beefing up the fleet of U.S.-flagged merchant ships that were built in U.S. shipyards and were owned and crewed by Americans. (This loosely parallels NASA's recent efforts to support emerging commercial cargo and crew services.) A key stimulus to the shipbuilding industry was the construction differential subsidy, which covered the difference in price between a ship built in the United States and a comparable ship built abroad. The stimulus for ship operators was the operating differential subsidy, which paid the difference in operating costs between an American operator, using an American ship and crew, and its foreign competitors. Under the statute, government loan guarantees were made available to purchasers of new ships. The Act also created the Maritime Commission, which initiated a program to build 50 new ships a year over the next decade. This program, although it didn't initially meet its quotas, helped give the shipbuilding industry a running start on the formidable challenges that it would have to meet during World War II.

During the war, U.S. industry produced nearly 5,000 ships. Despite wartime losses, this gave the U.S. far more shipping capacity after the war than it had before. As had been the case after World War I, the U.S. government sought to reduce its inventory of ships. This was implemented through the Merchant Ship Sales Act of 1946, which authorized the sale of cargo ships at attractive prices. Not all of the benefit went to the U.S. merchant marine fleet, however. More than half of the nearly 2,000 ships sold went to foreign interests.

More recently, the Merchant Marine Act of 1970 aimed once again to revitalize ship operators and builders. The goal was construction of 300 new merchant ships over a period of ten years. Other features of the Act were continuation of construction and operating differential subsidies, expansion of the federal ship mortgage insurance program, institution of a new capital construction fund program, and expansion of the maritime research and development program.

Implementation of the Act faltered as a result of the 1973 oil embargo by the Organization of Petroleum Exporting Countries and the recession that followed, which diminished worldwide demand for shipping.

Today, the full range of maritime activity from recreational boats to the largest ocean-going cargo and passenger ships are accommodated by about 360 commercial ports in the United States. These are governed by various state and local public port authorities, navigation districts, and municipal departments. Public port agencies manage the flow of waterborne commerce, but their mandate also includes promoting economic growth. They often build and maintain terminal facilities for passenger handling and for intermodal transfer of cargo—that is, the links that allow cargo to smoothly transition between ships and nearby rail and trucking services.

As noted, the U.S. government activity on behalf of shipbuilders and operators has parallels in space-related industries. The shift of space launch and satellite remote sensing services to the private sector began with legislative and executive branch actions in the early 1980s, and the legal and regulatory environment in each of these areas has continued to evolve. Both of these services are licensed by federal agencies, and in the case of launchers, that license brings with it some indemnification against catastrophic loss claims. Some of the companies enjoy access to government facilities and benefit from long-term government contracts. All have benefitted immensely from government research and development investments of decades past, and the emerging commercial human spaceflight industry continues to receive development assistance from NASA.

All U.S. launch providers are helped by a long-standing policy that requires U.S. government payloads to fly on U.S. launch vehicles. This policy may need to be reconsidered at some point. The U.S. government someday could find that its demand exceeds the supply of U.S. launchers, due to either a lack of qualified providers or diminished capacity caused by technical or business setbacks (e.g., a launch accident). Alternatively, the U.S. launch industry may find that the protectionist policy prompts other governments to respond in kind, preventing U.S. participation in lucrative launch contracts elsewhere in the world.

For a lesson on the unintended consequences of protectionist measures, the U.S. can look once again to its experience in the maritime industry. The requirements for designation as a U.S.-flagged vessel—for its crew, construction, and ownership—are among the most restrictive in the world.

Regulations for U.S.-flagged vessels engaged in trade require that all officers and pilots, as well as 75 percent of other onboard personnel, be U.S. citizens or residents. In addition, these vessels must be owned by U.S. citizens and constructed in U.S. shipyards. The construction requirement includes the hull and superstructure of the ship and the majority of materials used for outfitting the vessel. As a result of this restrictive environment, nearly 90 percent of the commercial vessels calling on U.S. ports are not U.S.-flagged. The regulations were driven by political considerations and designed to preserve entrenched businesses and organizations. Instead of protecting U.S. shipbuilders, operators, and related industries, the effect was to drive businesses to flags of convenience, slow the reforms needed to take advantage of containerized shipping, and undermine the competitiveness of the U.S.-flagged fleet. Policy-makers should take care not to make the same mistakes in planning the future regulatory environment for commercial spaceflight.

The public-private collaboration in maritime activity faced daunting challenges as the 20th century brought unprecedented development in ship propulsion, hull design and manufacture, cargo handling tools and techniques, communication and navigation capabilities, and perhaps most visibly, the size and specialization of ships. Existing and new ports had to be much larger and adopt new technologies and procedures to remain economically viable, which often meant moving to a new location and rebuilding from the ground up. Personnel needed new skills and different training in an environment where mechanization significantly reduced the size of the workforce. This rapid development, concurrent with the emergence of the modern international regulatory regime, makes this period of maritime history analogous to 21st century commercial spaceflight evolution.

Railroads: the first high-speed travel

In 1962, NASA provided a grant to the American Academy of Arts and Sciences to study the societal impact of the space program. The Academy chose to do this through historical analogy, with the development of American railroads as its focus. The result was the 1965 book *The Railroad and the Space Program*. Recognizing the risk of attributing too much predictive power to a single analogy, the book's editor treated the space program "not merely as an isolated matter, say, of exploration, or military preparedness, or scientific innovation, but as a complex *social invention*."

The railroad is a social invention that may give some clues as to how the development of space will progress. Its lessons, however, are sometimes misinterpreted by those who mistakenly believe that it was created entirely by business investors and served the nation only as a generator of economic profit.

Building networks, the old-fashioned way. In a sweeping history of the worldwide emergence and evolution of railroads, author Christian Wolmar credits railways with revolutionizing societies and economies in the 19th and early 20th centuries. Wolmar, a prominent transportation journalist in the U.K., sees the U.S. experience this way:

> It is the United States where the arrival of the railway in the nineteenth century had the greatest influence, both in creating the nation and in stimulating its economic development. Quite simply, without the railway, the United States would not be the United States of America.

Behind this exuberant assessment is more than a century of booming development, aided by massive government stimulus, encouragement of standardization in hardware and procedures, and regulatory intervention to improve safety in an industry that experienced far too many tragic accidents.

Just as commercial spaceflight hopes to do, rail transportation in mid-19th century America enhanced people's lifestyle by allowing them to experience things that were new to them. New possibilities for a broad swath of Americans included visits to cities and other points of interest far from where they lived; routine enjoyment of foods, household goods, and other products previously unavailable; much faster receipt of mail and news; the ability to sell local products to customers far and wide; and the opportunity to aspire to a career with the railroad or other industries made accessible by the railroad. In other words, the rail network can be seen as the "Internet revolution" of the 19th century. Its effects on people's lives were even more profound than those of the Internet today, because trains offered not just a better way of doing many things; for most people they offered the first and the only way of doing these things. They also had a technological "wow" factor, pulling loads greater than an entire wagon train at the amazing rate of up to 30 miles per hour—a speed at which, many believed at the time, humans would be unable to breathe.

In many parts of the world, rail systems had the effect of politically, economically, and socially uniting nations. This was both by accident and by design, and was characteristic of geographically small countries (such as Belgium, France, and Germany) as well as large ones (including Canada, Russia, and the U.S.).

We tend to think of the global proliferation of technology as a phenomenon that appeared in the late 20th century, but in the case of rail transport, it was alive and well much earlier. Between 1830—when the first fully operational passenger line between two major towns linked Liverpool and Manchester, England—and the end of the century, well over 600,000 miles of railway (including 200,000 miles in the U.S.) were built worldwide, touching almost every country.

Encouraging an industry this important to U.S. development was clearly an effort that the federal government had to undertake, and it did so in a variety of ways. Sometimes today we look back at the rapid growth of railroads in the 19th century, noting that private companies laid the track and owned and operated the trains, and we underestimate the level and importance of the government's role. In fact, a rail system of this scale would not have been possible without federal, state, and local governments acting in partnership with the private sector. Despite the high level of capital investment that private interests were able to apply, it wouldn't have been enough. In the 1840s, for example, more than 10 percent of the capitalization of the railroads came from state and local governments, and that increased to as much as 30 percent by the start of the Civil War. The famous transcontinental rail project was supported by federal loans amounting to $60 million (in 1860s dollars).

A fanciful depiction of the May 10, 1869 transcontinental railroad linkup at Promontory, Utah, from the May 29, 1869 issue of Frank Leslie's Illustrated Newspaper. The momentous event was made possible by many years of public-private collaboration, so the trains could have been labeled "Public Sector" and "Private Sector" instead of "San Francisco" and "New York." Source: Library of Congress, Prints & Photographs Division

Even more substantial were the land grants, which were more than just thin strips wide enough to lay the tracks. They included a considerable swath, extending a mile or more on each side of the tracks. Once the railroad and its depots were built, the rail companies sold the excess land for farming, lumber, and other businesses that sprang up in the towns along the way, some of which the railroad had created. By some estimates, the land grants to the rail industry in the 1800s were equivalent to the size of Texas, and the companies made a lot of money from them. One other seemingly insignificant but actually quite substantial freebie: most of the surveying of the land grants was done by the government and handed over to the rail companies.

The government did more than just provide resources and buy service to help jump-start the rail industry. Inevitably, as railways developed into national networks and crossed international borders, the need for standards for hardware, procedures, and regulations became evident. By the 1840s, many rail lines around the world were adopting the standard track gauge of 4 feet, 8.5 inches developed by British rail designer George Stephenson. This was helped by the fact that the British at the time were the dominant manufacturers of locomotives. Although it was becoming the standard, this gauge was far from universal. Narrower gauges offered lower capital expenditure, shorter turning radius, and less clearance needed for the train's width, which was particularly helpful when terrain clearance and tunneling were required. Wider gauges allowed greater capacity and stability. Thus emerged passenger trains with gauges ranging from about three feet to just over seven feet, hindering the linkage of rail networks across national borders, and sometimes even within national borders. For example, individual states in Australia competed with each other using multiple gauges, significantly slowing the advantages that should have ensued on a continent that is a perfect candidate for rail transport.

Track gauge was not the only standardization concern. Safety, efficiency, and ease of maintenance demanded conformity in boiler design, braking systems, coupler design, and track signaling systems. In today's space industries, the technologies may be different, but the importance of standardization remains. It's not a question of whether or not to standardize; rather, the questions are when and how.

Another long road to safety regulation. Train accidents became headline news in the U.S. starting in June 1831 when a railroad worker was killed in a boiler explosion on the South Carolina Railroad. From that point on, the deaths and injuries grew dramatically as rail services expanded across the country. There were hundreds of train accidents in 19th century America, many of which resulted in dozens of deaths and injuries, not just from exploding boilers but also from head-on collisions (often called "cornfield meets"), derailments, and bridge collapses as locomotives and railcars grew heavier. Newspapers published annual tallies of deaths and injuries from train wrecks, causing many in the population to shun rail travel.

Accident recordkeeping by the government and the industry was poor before the 1870s, when public outcry finally compelled Congress to order a government investigation of rail safety. Modern standards of accident reporting were not required until 1910, but even the information that was collected before that time presents an alarming picture of safety deficiencies across the railroad industry. Total casualties for the previous 10 years reported in the *Railroad Gazette* in 1893—the year of the first railroad safety legislation—included 5,623 deaths and 20,445 injuries.

About half the injuries and well over half of the deaths were railroad workers, making this one of the most hazardous industries in America. By the reckoning of the Interstate Commerce Commission (ICC), 44 percent of the worker casualties were incurred while coupling cars, which was done manually using nonstandard coupling systems. This safety problem was so prevalent that it was believed the experience level of a brakeman or switcher could be estimated by the number of fingers he was missing, and sellers of artificial limbs ran ads specifically directed at rail workers.

While many accidents could be blamed on poorly maintained boilers and user-unfriendly coupling systems, others resulted from inconsistent (or nonexistent) timekeeping, signaling systems, procedures, and training. As rail traffic increased, schedules remained erratic, even in places where trains shared a single track for both directions. Delays of various kinds were only part of the problem. A critical element that had to be standardized was time. England, which lies entirely within one time zone, standardized in 1847, but the U.S. didn't follow suit until November 1883. Before that, towns set their time based on their longitude, which meant that each state had many time "zones," sometimes numbering in the dozens.

Signaling systems became essential for traffic management, but in the days before electronic communications, visual signals were the only option. All operators in a region needed to be able to interpret the signals, regardless of which rail company they worked for. That forced cooperation among all users of the line as well as imposing training requirements on operators. In Europe, railway operators discovered that lack of coordination slowed cross-border traffic considerably, as crews needed to be changed at each border crossing, in part to ensure familiarity with local signaling and other technical systems, a problem that persists to some extent to this day.

In the closing decades of the 19th century, rail passenger-miles increased, technology continued to advance (including development of automatic couplers), public outcry for safety regulation grew louder, and the rail companies began to realize that federal regulation was preferable to conflicting state-by-state regulations. This opened the door for the first railroad safety legislation.

From a modern perspective, it seems remarkable that it took the United States six decades to begin regulating safety on railroads, during which time trains became the dominant mode of long-distance transportation, railroad companies became some of the nation's largest businesses, railroad work became the second-most dangerous occupation in the country (after coal mining), and the cumulative casualties numbered in the tens of thousands.

Federal regulation finally came in 1893 with the Safety Appliance Act, which was updated and expanded in 1903 and 1910.

The positive effects on worker safety were clearly evident even before the 1910 Act. In the 15 years following the 1893 Act, deaths among train workers due to coupling accidents dropped by a third, and injuries dropped by two-thirds. This is especially remarkable since it occurred during a time when the reporting of accidents improved and the number of train workers grew by a third—developments that elicit an expectation of *increased* accident totals. (Congress required the railroads to report accidents starting in 1901, but the authority to enforce the requirement with fines and to conduct investigations didn't come until that law was repealed and replaced by the Accident Reports Act of 1910.)

Another safety statute developed at this time was the Hours of Service Act, which the railroads fought vigorously in the years prior to its passage in 1907. The government made the case that extremely long workdays were contributing to accidents. The companies argued that limits on work hours would negatively impact their service and would be too costly. Few had sympathy for the companies' argument that the statutory limit was too onerous, and even fewer would today, since the legislation restricted rail employees' workdays to no more than 16 hours.

Boiler explosions, another major cause of injury and death, were addressed starting in 1911 by the Locomotive Inspection Act, which also was opposed by the railroad industry. Interstate Commerce Commission inspection authority in this area was expanded in 1915, and again in 1924 when inspections of diesel and electric locomotives became required in addition to steam engines.

Standardization of track signaling systems was a lengthy process that involved numerous studies and tests over many years. The results included implementation orders to railroads in the early 1920s and the Signal Inspection Act in 1937. Over time, numerous statutes have continued to refine the regulatory regime in areas such as inspection of tracks, rolling stock, and signaling systems; handling of hazardous materials; and employee alcohol and drug prohibition and testing.

Freight carriage has remained the backbone of cargo transport within the U.S., but passenger rail service went into decline in the mid-20th century and has never recovered. This decline was caused by several factors, but contrary to the beliefs of many armchair analysts, government involvement and the regulatory regime were not significant factors. Simply stated, it was the changing market environment. Passenger rail lost its dominance because of two developments: the proliferation of the automobile starting in the 1920s, accelerated by the post-war economic boom and the creation and improvement highway systems across the country; and the rapid growth of air travel in the decades following World War II. If the government deserves any blame for the demise of passenger rail, it's due to the extensive assistance it gave to competing modes of transportation.

Many in the rail industry didn't see this coming. Christian Wolmar quotes a remark made by a U.S. railroad manager in 1916: "The fad of automobile riding will gradually wear off and time will soon be here when a very large part of the people cease to think of automobile rides." Years later, passenger rail companies turned down offers of government aid to maintain their viability, not wanting to risk losing their commercial freedom.

Rail freight services did not succumb to competition because they provide an ideal and perhaps essential complement to other modes of freight hauling. Railroads offer a nationwide infrastructure capable of moving huge amounts of bulk, breakbulk (mixed), and containerized cargo across the country to nodes linked to the trucking, shipping, and air freight industries. Passenger rail, in contrast, is perceived by most of its potential user community as too slow and no less expensive than air travel.

There is a lesson here for commercial spaceflight similar to one we saw in the discussion of maritime transport. Rail became recognized as an essential service, for both passengers and cargo, soon after it was introduced. That allowed it to survive unregulated for decades despite countless accidents, many of them fatal. Today, rail cargo service is still considered essential, but rail passenger service is a shadow of its former self and needs government subsidies to survive. Commercial human spaceflight is under pressure to prove itself, and is not yet considered essential. If it achieves that status, the industry then must strive continuously to maintain it.

Roads to everywhere

Unlike the high-tech and revolutionary railways that sprung up in the 19th century, roads have been around since ancient times. Despite the perception of roads as old, proven technology, states and interest groups in the U.S. have been looking to the federal government for assistance in road building and maintenance since the nation's earliest days. In the 18th century, most roads were little more than paths, maintained by the residents, to connect farms to nearby towns. As president, George Washington was an advocate of road building, but it wasn't until Thomas Jefferson's administration that federal support for these efforts became established. Mainly, this involved sending funding to the states that was derived from the sale of public lands. The Jefferson administration also devised a plan for a network of federally funded roads that would connect cities as well as key shipping points along rivers and canals. Congress rejected the plan because it was seen as far more costly than the country could afford at a time when the federal budget was reliant on tariff revenues. Also, many believed that the Constitution needed to be amended to allow the federal government to take on projects of this type.

Jefferson settled for a less ambitious but still important project to build the Cumberland Road (also called The National Road), which provided an overland link between the Potomac and Ohio rivers. It was in bad shape by the 1820s due to heavy traffic and poor maintenance, so something needed to be done. Federal funding of road projects accelerated dramatically during Andrew Jackson's presidency (1829-1837), and during this time the states took on responsibility for maintenance, in many places instituting the practice of collecting tolls.

Toll roads, many of them run by private companies, had already begun to proliferate in the 1790s. Those that weren't on heavy traffic routes didn't make a profit, and most private operators were out of business by 1900. States, which were forced to take over abandoned toll roads lest they fall into disrepair, eventually created highway departments and commissions. By this time, the increasing popularity of bicycles had spawned the Good Roads Movement, which was reinforced by the growing acceptance of automobiles. There were 8,000 cars registered in the U.S. in 1900, and this number jumped to 33,000 just three years later. Over 13,000 vehicles were produced in the U.S. in 1911, and this grew to 56,000 by 1913.

Federal and state attention to roads didn't emerge just to enable auto and bike rides in the countryside for city folks on Sundays. As farming and manufacturing output grew around the country, there was increasing interest in improving the transport of goods to markets and shipping hubs. Better roads also helped extend daily postal delivery to outlying areas. The American Association of State Highway Officials was formed in 1914 to assist the federal government on relevant legislative, technical, and economic issues.

U.S. involvement in World War I, first as a supplier to European allies and later as a combatant, made the need for good roads particularly acute and heightened interest at the federal level. As the volume of activity exceeded the ability of the railroads to accommodate it, the trucking industry was born—an important private-sector development that would not have been feasible in the absence of the government-supported road system.

From that time on, legislative and bureaucratic interest in the expansion, maintenance, and safety of American roads became entrenched. In 1922 alone, more than 10,000 miles of federally funded highways were built. That signaled an impressive burst of activity, but there was a lot more improvement needed, with federal and state governments sharing funding, planning, and oversight duties while private industry assisted in standards development and provided construction, maintenance, and retail services that were essential to a functioning system. Major highway legislation appeared every few years in the following decades, often expanding the investment of federal resources.

It's common knowledge that the Interstate Highway System began construction in the 1950s under the Dwight Eisenhower administration. However, few are aware that the initial plan for that system was laid out in 1941 by a national highway committee appointed by President Franklin Roosevelt, and the idea had been proposed as early as 1913. It took decades for everything to come together: justification for federal investment in such a big project, agreement on a plan for the road network that would satisfy the stakeholders at the state level, and appropriation of funding to get the project started and then carry it through to completion.

Eisenhower used national defense as part of the justification for the highway system, judging that it would allow more efficient movement of personnel and equipment around the country and provide long stretches of straight roads that could be used to land aircraft in an emergency. But the growth in traffic offered all the justification that was necessary. Vehicle registrations jumped to 49 million by 1950, a 60 percent increase in the five years since the end of the war, and reached almost 63 million by 1956.

The first appropriation of funds for Interstate Highway System construction came as a result of the Federal-Aid Highway Act of 1952, but this amounted to just $25 million in startup money. The project's future wasn't guaranteed until the Federal-Aid Highway Act of 1956 pledged billions of dollars over the following three years, to be paid for by increased taxes on fuel and tires and new taxes on trucks and buses. An important innovation in the legislation was the establishment of the Highway Trust Fund, which would guarantee that the taxes collected would be set aside for highway projects.

It almost didn't happen. When the bill was first introduced in 1955, it was defeated in the wake of lobbying by the trucking industry, which objected to the new taxes and had the support of the oil and rubber industries. By the next year, the truckers realized that their maneuver was counterproductive—if they didn't pay the taxes, they wouldn't get the highways, and the future of their industry would be shortchanged. A new version of the bill passed overwhelmingly in 1956.

About 12,500 miles of interstate were open by mid-1962, with an average of 34 new miles added every week. Thankfully, as the interstates grew, safety improved. Despite the huge increase in the number of vehicles, the fatality rate on U.S. roads dropped from 16 per 100 million miles traveled in 1929 to just 5.5 per 100 million miles in the early 1970s. And fatalities on the interstates occurred at less than half that rate.

By the end of the 20th century, there were 200 million vehicles on American roads, and the Interstate Highway System had grown to about 50,000 miles. The overall benefit to society and the economy from federal investment in roads would be impossible to accurately quantify. The collaborative endeavor with state and local governments and the private sector, despite occasional complaints from the user community about taxation and regulation, paid off handsomely. Collective efforts across several decades produced an extensive, sophisticated road system that many observers had thought was too big a challenge—not unlike the future space infrastructure that will take shape in the coming decades.

Aviation: spaceflight's older brother?

It's often taken for granted that aviation is the best analogy for space. The link is understandable since both activities involve flight, and their communities often overlap in large corporations and government agencies, thus contributing to the origin of the word "aerospace" early in the space age. Aviation and space share other characteristics as well: both have great utility in national security and economic pursuits, and both have been highly esteemed for their contributions to national prestige and human imagination, which have helped them overcome their sometimes less-than-favorable economic merits.

Nonetheless, at times too much is made of this comparison, as when observers express their annoyance that space transportation has not evolved as quickly as air travel—a notion that ignores the difference in the degree of difficulty. We wouldn't think of comparing, for example, aviation's development with that of consumer electronics. If we did, we'd be asking absurd questions such as, "If the electronics industry can make the dramatic advances in speed, efficiency, and cost reduction that we've seen in recent decades, why can't Boeing make a commercial airliner that costs $500 and circles the globe in 20 minutes on five gallons of fuel?" The answer, of course, is that atmospheric flight confronts the limitations of physics much more quickly than electrons traveling through an integrated circuit, so the analogy is inappropriate. Such is the case for aviation and space when we talk about the pace of development cycles and the magnitude of advances.

The paragraphs that follow provide a brief glimpse at the aviation side of the story, which holds lessons for current and future space development. Not in the pace or direction that spaceflight could or should take, but rather in what happens in the evolution of big infrastructure projects. Such projects demand extensive research and development efforts, have significant public policy implications, and can benefit from a complementary relationship between the public and private sectors. This relationship has experienced some episodes that weren't positive and productive, but as technology historian and analyst Tom Heppenheimer noted in his history of commercial aviation, the industry "has benefited enormously from policies of government... The U.S. Air Force brought forth the jet airliner, and went on to lay the groundwork for today's wide-bodies, including the Boeing 747. Governments also promoted air safety by taking responsibility for air traffic control."

The early years. The public-private collaboration got off to a slow start. Scientist and inventor Samuel Langley, who was believed to have America's best shot at creating a working airplane, received a $50,000 grant from the U.S. War Department, plus another $20,000 from the Smithsonian Institution, in 1898. Despite his valiant efforts, he failed to produce a powered flying machine that would carry a pilot. The Wright brothers beat him to it in 1903 without the benefit of government funding. But the ultimate success of their invention depended on selling it, and the government was the most obvious potential customer. The U.S. Army purchased a Wright B Flyer in 1909, procured under a letter contract that stipulated a price, a delivery schedule, and minimum performance requirements. The Army experimented with the Wright B Flyer from 1909 to 1912, but never ordered a large production run.

As Orville Wright was demonstrating the aircraft to the Army in fall 1908, Wilber Wright went on a demonstration and sales trip in Europe. He found keen interest there in an aviation community that was beginning to flourish, and would continue to do so in anticipation of the war that was to come in the following decade. By 1910, the Europeans had surpassed the

U.S. lead in aircraft technology and controlled flight, helped by the use of public-private partnerships to pool expertise and funding. While most of the European combatants fielded aircraft in World War I, the U.S. had no aircraft in combat.

Recognizing that the U.S. was falling behind, and spurred by the outbreak of war in Europe in 1914, the National Advisory Committee for Aeronautics (NACA) was created in 1915. Initially an effort to simply coordinate existing aviation activities of the Army, Navy, Smithsonian, and other U.S. organizations, NACA began operating its first laboratory, named for Samuel Langley, in 1920. The decades that followed saw U.S. aviation keep pace with and eventually exceed that of Europe, thanks in part to work done by NACA.

NACA research mainly focused on applied engineering related to the design and testing of airfoils, airframes, and engines; improvements such as control systems, retractable landing gear, and metallurgical techniques; and some theoretical work, especially in aerodynamics. NACA also had projects that directly addressed safety, such as icing studies and engine-out flight tests of new airliners as they prepared to enter service.

NACA was a critical participant in another important activity during the years of the first World War that was unrelated to engineering research: it played the role of mediator in the intellectual property dispute that erupted between the Wright brothers and aviation pioneer Glenn Curtiss. The Wrights believed that their 1906 patent, which described wing warping and mentioned an aileron-type control system, entitled them to royalties for any revenue-generating use of lateral control devices on aircraft. This was an exceedingly broad claim. Without lateral control—basically, the ability to roll to make turns and maintain stability—an aircraft would be practically useless, so all future aircraft could be expected to employ such devices. The Aerial Experiment Association, created by inventor Alexander Graham Bell in 1907, employed ailerons in its designs, but the organization disbanded after building several planes. Curtiss, one of its founding members, parlayed this experience into a business venture. On behalf of his company, he flew an aileron-equipped plane in front of a paying audience in June 1909, prompting a lawsuit from the Wrights.

Although the Curtiss-Wright case had the most visibility, it was not the only instance in which patent concerns threatened to slow U.S. innovation. As the nation entered World War I, the government realized that acquiring airplanes would be difficult and expensive because essential patents were dispersed across several holders who were charging exorbitant royalties. Many patent violation lawsuits were pending among airplane manufacturers, nearly halting production of aircraft in the United States. Two developments in March 1917 put this problem on a path to resolution: an advisory panel chaired by Franklin Roosevelt (then Assistant Secretary of the Navy) recommended forming a patent pool, and the Congress passed the fiscal year 1918 Naval Appropriation Act with $1 million set aside for purchase of airline patents. Along with the NACA intervention, this led to the formation of the Manufacturers Aircraft Association (MAA). All major aircraft manufacturers became members of MAA and agreed to waive all claims against each other for actions prior to July 1, 1917. MAA established an arbitration procedure for future claims.

The U.S. government didn't need to spend the $1 million appropriation to buy patents because a cross-licensing agreement was worked out in a series of meetings arranged by a Subcommittee on Patents set up by NACA. The agreement solved a big problem, but created another. Opponents of the agreement saw it as an illegal trust designed to favor large, established players and discourage new entrants. Throughout its existence, MAA was attacked by the press and some members of Congress, although numerous congressional investigations failed to find fault with its policies and procedures. The government filed an anti-trust suit against MAA in 1965, and the long legal battle resulted in the association's termination in 1977.

Although MAA didn't please everybody and was eventually disbanded, its activity and the government's encouragement overcame a hurdle that could have undermined U.S. aviation efforts for an extended period, putting the nation far behind other parts of the world where public-private partnerships were designed to circumvent roadblocks and accelerate development. Patent disputes may seem like mundane concerns that are peripheral to building and flying aircraft (or spaceships), but such things can make or break companies and sometimes whole industries. A space-related example is the long-standing arrangement of cross-waivers of liability for NASA launches. Rocket launches that go bad have the potential to be very messy affairs, causing considerable injury, loss of life, and property damage. If big lawsuits are the likely result, companies will be disinclined to be launch providers for the nation's civil space program. To encourage participation, NASA and its contractors and subcontractors mutually agree not to sue each other in the event of launch mishaps, except in the case of gross negligence. Through government-industry cooperation, the risk is managed and the activity can proceed.

Regulation and subsidy help bring commercial viability. The successful emergence of the aviation industry served both public and private-sector interests. Like the automobile industry of the time, early 20th century aviation companies were started by a new breed of entrepreneurs and visionaries who were producing a wide variety of aircraft by the early 1920s. Barnstorming was in full swing, providing adventurous civilians—perhaps unaware of the risks, or at least willing to overlook them—a chance to take flight. Eventually, the budding aviation industry would contribute more to the nation's economy and military strength than anyone at the time could imagine.

The evolution of flight safety regulation and practice was accelerated compared to rail and shipping, in part because those older industries laid the groundwork for federal government oversight and enforcement and had shaped public expectations for safety and reliability. Steam-powered shipping in the U.S. saw a gap of about 30 years between its first commercial service and the first safety legislation, and for rail the gap was over 60 years. In both cases, credible monitoring and enforcement mechanisms took many additional years. The first commercial services in aviation appeared in the 1910s and regulatory statutes began to appear just a decade later, followed quickly by serious efforts at implementation.

The commercial operators of rail and shipping enterprises had fought strenuously for decades to prevent any safety regulation from interfering with their business, and often tried to circumvent the regulations once they were adopted. The difference in aviation was that industry hopefuls initially encouraged regulation to improve their business case. As rail and shipping discovered the hard way, safety and reliability (and importantly, the perceptions of customers and investors) benefit from standardization and precautionary requirements. Consistent rules across the nation, greater public confidence, reduced liabilities, lower insurance costs, and improved ability to attract investors were essential to the success of commercial aviation.

The U.S. Post Office ran its own Air Mail Service from 1918 to 1927, following strict maintenance and inspection procedures and insisting on proper training for pilots and mechanics. Despite this, in its first three years the service experienced 27 fatalities in 1.5 million flight miles largely due to a management philosophy that "the mail must go through" which required pilots to fly on schedule regardless of weather conditions. When that philosophy was changed to allow regional managers and the pilots themselves to determine when it was safe to fly, the casualty rate dropped dramatically. For the remaining years of the service, there were 16 fatalities in 12.25 million flight miles, giving it a fatality rate that was better than that of the emerging (and unregulated) aviation industry by a factor of 60.

The Kelly Air Mail Act of 1925 required the Post Office to give up its air fleet and turn the responsibility over to private contractors, which it did by soliciting bids on individual intercity routes over the next two years. The Ford Motor Company was the first beneficiary of air mail contracts, followed quickly by the Boeing Company. Contractors were encouraged to carry passengers as well—indeed, they were compelled to seek passenger revenues by changes in the air mail rate structure after 1930—helping to jump-start the airline industry. (In 1927, the first year of its air mail contract, Boeing carried 525 passengers.) This encouragement was necessary to overcome the industry's reluctance to carry passengers due to perceived inability to provide adequate comfort and safety, and lack of profitability. There was more enthusiasm among European carriers at the time for passenger services. European governments were more forthcoming with subsidies, and wanted to establish connections across the English Channel and to colonies as far away as South Asia and the East Indies. Similar incentives did not exist in the United States, with its contiguous landmass connected by a thriving passenger rail system.

The expected rapid expansion of commercial aviation in the wake of the Kelly Act raised concerns about industry's poor safety record compared to the government-run Air Mail Service. Regulation became essential, and was soon mandated in the Air Commerce Act of 1926. The Act established the Aeronautics Branch in the Department of Commerce (DoC) to handle licensing of airmen, certification of aircraft, and development and enforcement of air traffic rules. The Branch's Air Regulations Division was responsible for safety.

In order to keep costs and workloads manageable for both government and industry, the Aeronautics Branch instituted two practices. First, medical exams of airmen would be conducted by designated doctors in private practice, at airmen's expense, rather than employing a government medical staff. Second, aircraft would be certified through type-certification, in which each model of aircraft would be inspected and tested for adherence to minimum standards to earn airworthiness certification for all aircraft of that model, thus avoiding the burden of certifying each individual aircraft. Despite these practices, the growing industry quickly produced backlogs that did not clear up until the Great Depression slowed development.

The aviation community expanded rapidly, and regulatory oversight was challenged to expand with it. Starting in 1930, aviation businesses—airlines and flight schools—were required to obtain certification along with pilots, mechanics, and aircraft. In 1935, the Bureau of Air Commerce (the new name of the Aeronautics Branch) added certification requirements for airport controllers.

In the 10 years after the Air Commerce Act, the U.S. aviation industry grew by a factor of several hundred as measured by the number of pilots and passengers and the amount of revenue generated. At the same time, the rate of fatal accidents per aircraft mile went down dramatically. But safety regulations and their implementation continued to evolve as the industry grew further, technologies changed, and high-profile accidents persisted.

The remarkable growth of the U.S. airline industry after World War II clearly called for increases in regulatory and inspection personnel and improvements in airport and air traffic control infrastructure. The Civil Aeronautics Act of 1938 had lifted a ban on federal support to airports, clearing the way for much-needed government investments such as those that followed the Federal Airways Act of 1946. Also in the late 1930s, the government made its first forays into operating the air traffic control system. This responsibility had been handled by the airports and airlines in an increasingly dysfunctional system hindered by awkward lines of communication, resistance to sharing information and best practices, and underinvestment in new technology. After a few years, it became evident that air traffic control was one of the areas where the government could do the job better and more efficiently than the private sector.

Transition to passenger jets. The number of airline passengers in the first half of the 1950s was more than double the number from the latter half of the 1940s. In the second half of the 1950s, the number almost doubled again, partly due to the introduction of jet airliners in 1958—the most visible, but far from the only technological advance that fueled growth throughout this period. As jet travel became entrenched in the 1960s, the number of U.S. passengers nearly quadrupled from the previous decade. An illustration of the surging capabilities that made this possible can be found in the evolution of Douglas Aircraft, which introduced the DC-3 in 1935 to carry 21 passengers up to about 1,000 miles at 165 miles per hour. Thirty years later, the DC-9 carried 120 passengers the same distance at 565 miles per hour.

Despite this growth, safety and regulatory resources lagged as administrations (particularly under Presidents Eisenhower and Johnson) and the Congress sought to trim spending, and aviation regulators were not spared from budget cutbacks. Meanwhile, regulators were called upon to respond to a number of high-visibility accidents, such as the in-flight collision of TWA and United airliners over the Grand Canyon in 1956, which resulted in 128 fatalities and put a spotlight on problems with air traffic control, and a December 1960 mid-air collision over New York—again involving TWA and United airliners—that killed 128 people on the planes and eight people on the ground. Regulators also needed to certify and monitor the new jet aircraft that were entering commercial service, introducing larger and faster planes that required expansion of airport facilities and close observance of their failure modes or other possible safety shortfalls.

By this time, industry's attitude toward regulation resembled the adversarial relationship historically evident in rail and shipping: new regulations were resisted, derided as too complicated, too costly, too intrusive, and unnecessary. For example, the Airline Navigators Association declared Doppler radar to be unsafe, and airlines opposed the requirement for flight data recorders and cockpit voice recorders due to cost, while pilots considered the voice recorders to be an invasion of privacy. Long gone was the 1920s view that prompted industry to request regulation to aid its long-term development.

By the late 1960s, the critical safety issues being addressed by the Federal Aviation Administration (FAA) included transport of hazardous materials, crashworthiness of airframes, and prevention of fires and explosions. At the time, a large percentage of airline fatalities resulted from post-crash fires.

Deregulation. The FAA's licensing, certification, and inspection duties became even more challenging as a result of the Airline Deregulation Act of 1978. The number of commercial carriers doubled between 1979 and 1983, accompanied by increased use of leased aircraft and outside contractors for maintenance, all of which complicated the inspection process. The increased competition raised concerns that some carriers would skimp on maintenance to control costs. Also, schedule pressure associated with the hub-and-spoke system provided incentive to cut corners. The FAA responded with efforts to streamline paperwork and improve the standardization and efficiency of its regional offices. Meanwhile, work continued through the 1980s on new rules for fire safety and airliner evacuation.

The deregulated environment continued to be a challenge in the 1990s, as demonstrated very visibly by the ValuJet crash in Florida in May 1996, which killed all on board. ValuJet had been in operation for less than three years, but in that short time it had grown to a fleet of over 50 planes. Its business model consisted of buying older, highly depreciated planes (its fleet was made up of DC-9s) and outsourcing their heavy maintenance requirements. In the wake of the accident, the FAA conducted a 90-Day Safety Review, which recommended "surveillance improvement," including increasing attention on new carriers in their first few years, seeking proof that companies have the necessary infrastructure and capabilities to expand their operations, and in general targeting trouble spots more diligently. By the end of the 1990s, these procedures were incorporated into the Air Transportation Oversight System that is active today.

From our perspective today, it seems strange that until the mid-1930s, the U.S. government had to cajole and subsidize the airline operators to take on passengers in order to stimulate a new industry that would be beneficial to the nation's economy. The airlines' reluctance stemmed from the simple fact that air passenger service didn't become profitable until the arrival of larger, safer aircraft that enabled more economical operations. (The DC-3, able to carry 21 passengers, was the first to achieve this.) In contrast, commercial human spaceflight is taking a different approach. The government has no need to coax commercial space entrepreneurs to seek passengers; they are eager to do so even though the profitability of the business is far from assured and may not be demonstrated for many years to come. Rather, the government is debating within itself whether it's appropriate to give the industry a helping hand, and if so, how much. In any case, the government has a role to play due to its responsibility for public safety and its commitments to international agreements that make it responsible for the space activities of citizens and companies under its jurisdiction.

One of the reasons for highlighting the public-private collaboration in aviation history is to dispel a myth I've heard people repeat many times over the years. The argument goes like this: "Aviation has grown tremendously and proven itself economically, but spaceflight hasn't, because aviation is a private sector activity and spaceflight is a government activity." It should be evident by now that this is complete nonsense, an attempt to rewrite history to suit a particular worldview. As this brief discussion of U.S. commercial aviation shows, the U.S. government and private sector have been closely intertwined throughout the history of powered flight. There is every indication that the same will be true for commercial spaceflight. While the relationship has been rocky at times, the net result for commercial aviation has been very positive, enabling and encouraging dramatic growth while demonstrating equally dramatic improvements in safety. In this way, the aviation experience resembles that of rail and maritime, although at an accelerated pace. If the spaceflight industry learns from its predecessors, and recognizes the expectations that have been planted in the public mind, it can improve its chances of a positive outcome at an even faster pace—one or two decades instead of several.

Learning lessons for spaceflight

In this chapter, we've seen the U.S. government act as a key player in infrastructure and commerce by creating and improving technologies, giving out land grants, formulating safety regulations, and improving the stability and predictability of the business environment by establishing regulatory uniformity across all the states. We've seen that in the absence of regulation, operators are motivated to cut corners to gain advantages in cost, capacity, and speed to the detriment of safety for passengers, workers, and third parties. We've seen industry eagerly accept government largesse and protection, but strenuously fight regulations, delaying their initiation by many years because they saw them as increasing their costs and hindering their flexibility. Regulations undoubtedly presented short-term challenges to businesses, especially after enforcement became more robust. But the long-term operational and cost benefits of standardization, coupled with improved safety performance, undoubtedly helped accelerate, rather than hinder, sustainable growth.

Rail and shipping blazed the trail in U.S. transportation safety regulation, each in its own evolutionary process lasting nearly two centuries so far. In both cases, the evolution of a comprehensive body of safety regulations was a century-long process during a period of rapid technological change, expansion of the user community, and numerous casualties. Aviation was a relative latecomer, but still offers over eight decades of experience. The spaceflight industry must learn quickly from its predecessors. The industry would like to experience comparable technical and market development in the coming decades (preferably without the casualties), but given the precedents set by the more established industries, it cannot expect to be granted such a generous timeframe in which to institute the rules and practices that will drive the performance and instill the confidence so essential to building a successful business. It wouldn't be surprising for a potential investor approached by a hopeful commercial spaceflight company to respond, "Come back after your industry's regulatory regime is in place and there's a track record showing that it works." Government regulators and the spaceflight industry should work together on this, trying to break the pattern of the rail, maritime, and aviation industries, in which major strides in safety came only in the wake of considerable loss of life.

Part 3
The Future

Chapter 8

Finding a path to the mainstream

> *Planning a human spaceflight program should start with agreement about the goals to be accomplished by that program—that is, agreement about its* raison d'être, *not about which object in space to visit. Too often in the past, planning the human spaceflight program has begun with "where" rather than "why."*—Augustine Committee, 2009

Learning from spaceflight history

From childhood, we're taught that it's important to study history so we can avoid repeating the mistakes of the past. As the previous chapter demonstrated, there's plenty of valuable experience to draw from, providing us with lessons that can be applied to current problems. Unfortunately, even those who study history don't always learn the *right* lessons from it.

The most basic mistake that people make when they look back at the Apollo era is to interpret President Kennedy's 1961 call for a manned lunar landing as a destination-driven strategy aimed at opening the Moon, and eventually the rest of the solar system, to permanent human activity. Actually, it was never intended as such. To some people, that may sound counterintuitive, or even blasphemous. After all, Kennedy told a joint session of Congress that "this nation should commit itself to achieving the goal, before this decade is out, of landing a man on the Moon and returning him safely to the Earth." He also spoke of space as "this new ocean" and pledged that the United States "does not intend to founder in the backwash of the coming age of space." That rhetoric sure sounds destination-driven and forward-looking, but in fact the intent behind it was to build capabilities and serve near-to-medium-term political needs. It is well documented that Kennedy's goals included boosting U.S. technological prowess and industrial production, demonstrating superiority over the nation's Soviet adversary, and winning over hearts and minds in non-aligned nations. One of the many occasions on which he made this clear was in an April 1961 memo to Vice President Lyndon Johnson which asked:

> Do we have a chance of beating the Soviets by putting a laboratory in space, or by a trip around the moon, or by a rocket to land on the moon, or by a rocket to go to the moon and back with a man? Is there any other space program which promises dramatic results in which we could win?

If Kennedy had been interested in the Moon as a strategic destination, his plan would have included some type of enduring lunar habitation that involved a mix of scientific research, resource exploitation, and possibly permanent settlement. But Kennedy was not a space settlement enthusiast—indeed, he was not a space enthusiast at all, aside from recognizing what it could do to further other national goals. Establishing a permanent, productive presence in space was never part of the Apollo program, and the infrastructure created to carry out the lunar landing was not designed to support such an eventuality.

Historical and political analysts who have been willing to take a dispassionate look at the early years of spaceflight all have concluded the same thing: Apollo happened at an anomalous time during which all the ingredients to make it possible—geopolitical, technological, economic, cultural—came together in ways that we cannot expect to recur. The message is clear: to base U.S. civil space policy and strategy on an expectation of such a recurrence would be folly. Apollo cannot be the definitive model for the things we do today, the choices we make for the future, or the pace at which we pursue them. Yet so many people at all levels of the debate still fall into this trap.

The end of the Cold War at the beginning of the 1990s should have been a wake-up call that it was time to rethink the justification and strategy for human spaceflight. But most people missed the call. Some of those who recognized it perceived it as a death knell. A colleague who was intimately involved in high-level space policy-making at that time opined recently that U.S. human spaceflight died at the end of the Cold War. Apparently, in this view, it's been going on momentum since then simply due to the space shuttle and space station programs.

What changed for human spaceflight at the end of the Cold War? Much has been said and written on this topic, so let me boil it down. During the Cold War, human spaceflight fulfilled two major national purposes: 1) it contributed to prestige, and 2) it was a research and development project. (In a few specific locations around the country, it was also a jobs program, and remains so today, as previous chapters illustrated.) At the time, this was sufficient to justify the investment and the risk. After the Soviet threat disappeared, these purposes still existed, but were *no longer sufficient* to justify indefinite continuation of the program in the United States—although they may still be sufficient for emerging spacefaring nations such as China and India.

That raises the question being asked all around the space community, especially among policy-makers in Washington: Whither the U.S. human spaceflight program? (Okay, I'm paraphrasing. Nobody in the space community actually says "whither.") This question may be intended to guide our strategic planning, but too often it leads to pitfalls. The discussion quickly devolves into a shopping list of solar system destinations that we think astronauts will be able to visit within the next decade or two. We then pretend that the time, resources, and risk involved in the effort can be justified by the science results, the aerospace jobs created, and the inspiration it will give to our youth. The goal of purposeful, sustainable space activities gets lost, as does the realization that creation of a diverse and thriving in-space infrastructure will be a multi-generational and multi-national effort.

As noted in the Augustine Committee quote at the beginning of this chapter, we should start by agreeing on goals, but we tend to begin with "where" rather than "why." This observation is a critical unlearned lesson that deserved to be displayed in flashing red letters on the first page of the Augustine Committee's report. Unfortunately, it was tucked away on page 111.

Where do we start?

Creating a productive future in space exploration and development requires difficult technical, economic, and political trade-offs. Crafting the right balance is critical, but complicated. We want to advance our technologies and capabilities, but there is disagreement on which approach offers the best chance of success: tying technical developments to specific missions, or more general investment in improving the state of the art in particular areas. A specific mission presents a clear goal, well-defined requirements, and often a firm schedule, but it's inflexible and doesn't leave room for general experimentation that could produce unpredictable advances. The final product may be a system with very narrow applicability that quickly gets discarded even if it's highly successful (a prime example being the Saturn 5 launch vehicle). On the other hand, too much reliance on undirected research efforts could be deficient in usable results as the quest continues for even better, more advanced, more elegant outcomes. Striking the right balance between the two approaches is a subjective exercise, more art than science.

Ideally, we'd like to establish a series of significant milestones, and achieve them as quickly as possible. That keeps the workforce busy, attracts fresh talent to the effort, and demonstrates accomplishments that help sustain public and political support. But pushing programs onto a fast track without good reason and adequate funding could be their undoing. Exaggerated promises and unmet milestones will have negative repercussions. Even if there is short-term gain, a hastily executed program could undermine long-term utility. There needs to be an assessment process that balances these many considerations and is widely viewed as credible.

The NASA Authorization Act of 2010 directed the agency to initiate a review of the strategy for human spaceflight in a process that has been compared to the highly successful National Research Council "decadal surveys" that are done for planetary science, Earth science, astronomy and astrophysics, and biological and physical space science research. As of this writing, the review hasn't yet begun. It's likely to be far more difficult than the scientific surveys, and may not produce results that are nearly as useful. Instead of drawing panel members from a fairly limited set of disciplines that are all aimed at increasing scientific knowledge, any attempt to assemble a human spaceflight panel inevitably brings demands for representation from a broad array of participants across the government, industry, academic, and advocacy communities. Multiple agendas from these sectors—seeking scientific discovery, engineering advances, bureaucratic survival, job security, benefits for political constituents, corporate revenues, and fulfillment of long-standing spaceflight dreams—could result in an inability to reach consensus on recommendations that don't conflict with each other and don't break NASA's budget several times over.

Human spaceflight systems development and operations tend to span far more than a decade, so it's not obvious how frequently a survey of this type should be revisited, or if it should be an ongoing strategic planning and oversight function. In any case, the analysis will be politically charged because national prestige and workforce issues will be prominent considerations, in contrast to the science surveys which are driven mostly by fundamental scientific questions.

The existing decadal surveys aren't followed to the letter, but they do provide guidance to their relevant communities that is respected and heeded as much as possible given the budgets available. This would be harder to do with a human spaceflight survey because these activities are by far the most expensive and are among the longest-lived projects that NASA undertakes. Also, devising a human spaceflight strategy in isolation from everything else doesn't make sense. In effect, such an approach says: "Assume a human spaceflight program. What should we do with it? Where should we send astronauts that will keep everyone interested?" This is a good way to make the endeavor look like a stunt or an athletic competition in which a crew reaches a "finish line." Once that happens, everyone's interest quickly wanes, crippling follow-on efforts.

If a high-level survey has questionable utility, how do we begin the process of setting sustainable long-term goals and crafting a strategy to carry them out? The first question that should be on the lips of policy-makers is: "What program of space exploration and development would best serve U.S. national interests?" Then, *after* a reasonable answer to that question has been developed, the next question is: "What is the role of human spaceflight in this program?"

Before setting dates for human missions to asteroids or Mars, the U.S. needs to define the purpose of such missions as part of a larger strategic plan in the national interest. A plan of great scope and duration is extraordinarily difficult to formulate, gain approval for, and sustain. But enduring success in a resource-constrained environment demands that we undertake the difficult process of formulating and winning approval for a plan to expand human and robotic activity throughout the Earth-Moon system and then to other parts of the solar system. This can't work if we try to substitute tools and tactics ("Let's build a new rocket that can fly beyond LEO") for goals and strategies ("Let's create new knowledge, new capabilities, and long-term benefits"). If we make this mistake, we risk expending too much of our resources on designing, building, and testing hardware with little left to operate it or develop payloads for it, thus defeating the purpose of the whole exercise.

Spaceflight in the mainstream

Some space activities have gone mainstream. Certainly this has been true for a long time with communications and weather satellites, and more recently with navigation satellites. Other types of satellite remote sensing beyond weather monitoring have been used routinely for decades by the scientific and military communities, and are now commonplace for the rest of us in areas such as news reporting and online mapping services. In contrast, human spaceflight and other space industries have not yet become mainstream and still face multiple challenges. Spaceflight is not a technological capability that the population grew up personally experiencing, like air travel or international phone calls. It's considered a "special" activity, and a very complicated one, so each generation of policy-makers, financial managers, investors, and the general public needs to be educated anew by space practitioners.

The problem is illustrated by the debate over NASA's budget and priorities described in previous chapters. One of the primary reasons for the existence of the space agency is to advance the state of the art in space technology. Yet when the Obama administration sought to reverse the decline of this activity that has prevailed in recent decades, the proposal was assailed by opponents who saw it as a low priority, or even wasteful. This reaction, if it's sincere and not just partisan nay-saying, is astonishing in a global environment where space-related technologies are evolving rapidly and proliferating. It conveys a message that the U.S. is content to eat its seed corn while others plan for their next harvest. Space practitioners felt compelled to beg congressional appropriators to reconsider their opposition in a letter to NASA's Senate Appropriations subcommittee on September 8, 2011. The 45 companies and universities that signed the letter saw the need to remind the Congress of the high payoffs for the economy, education, quality of life, and future NASA missions, and the consequences of underfunding the Space Technology Program. Federally funded programs inevitably must address disagreements at the margins regarding their research agendas, but unlike mainstream pursuits such as medical research, space research is often called upon to justify its inherent worthiness for federal funding.

This is not to imply that the private sector is absolved from responsibility for investment in space research and infrastructure development. Industry must hold up its end of the partnership, as illustrated in the historical examples of the previous chapter. The mark of maturity in spaceflight will be the government's ability to rely on the private sector for establishment, ownership, and ongoing operation of significant portions of the space infrastructure. Earth-to-orbit transportation will be an early component of this evolution, but it will not be the only one. The question is, will the U.S. lead the way in this advancement, or will some other nation do it first?

CHAPTER 9

In search of forward-looking space policy

In the modern era, in which the goals of space exploration have expanded beyond a single target, the necessary technological developments have become less clear, and more effort is required to evaluate the best path for a forward-looking technology development program. NASA has now entered a transitional stage, moving from the past era in which desirable technological goals were evident to all, to one in which careful choices among many conflicting alternatives must be made.
—National Research Council, Steering Committee for the NASA Technology Roadmaps, 2011

In a groundbreaking book way back in 1929, Slovenian engineer Hermann Noordung called cislunar space, the area encompassed by the Earth and the Moon, "our immediate homeland in the universe." It's important to keep this in mind in our strategic planning for space exploration and development.

Think of cislunar space as analogous to the environment where you grew up. The Earth is the residential neighborhood of your youth, the Moon is the local elementary school and the playground where you learned to climb and swing and play in the sandbox with the other kids, and the space in between holds the local businesses, roads, and bus stops. In the early years of our existence, this is what we experience as the training ground for the rest of our lives. Moving into adolescence, we get our first job and start to range a little bit more widely—maybe across town, but not yet across the continent or the ocean. Our adolescence in space will see cislunar space developed into an industrial park where we can hold down permanent jobs, harvest materials and energy, and create new value-generating enterprises. When we reach adulthood, we can settle far from our original home, start a family, and use our wisdom and maturity to accomplish things we never thought we could do when we were younger. Adulthood in spaceflight means settlement in and beyond cislunar space, discovering and creating things that we couldn't even imagine as infants. That's what we are today—infant spacefarers.

This conceptualization of spacefaring evolution is akin to one expressed by the Space Task Group, appointed by President Nixon in 1969 to propose what should follow the Apollo program. They envisioned a long-range goal of human planetary exploration proceeding in three phases:

- Initially, activity should concentrate upon the dual theme of exploitation of existing capability and development of new capability, maintaining program balance within available resources.
- Second, an operational phase in which new capability and new systems would be utilized in Earth-moon space with groups of men living and working in this environment for extended periods of time. Continued exploitation of science and applications would be emphasized, making greater use of man or man-attendance as a result of anticipated lowered costs for these operations.
- Finally, manned exploration missions out of Earth-moon space, building upon the experience of the earlier two phases.

Like the stages of a human life, the three stages of space development—let's call them training ground, industrial park, and settlement—can't be reshuffled or skipped over. We have some control over the rate at which we traverse them based on factors such as the depth and persistence of our commitment and the level of our investment. Currently, these factors don't seem to be working in favor of rapid space development. But that's not an excuse to skip steps.

The slow evolution of U.S. space policy

Policy evolution—in general, not just for space—tends to lag behind changing socio-economic circumstances, market developments, and technological advances. That's because the impetus for change comes from a large and diverse community where every faction seems to be pulling in a different direction. The only time we seem to be able to move quickly is when there's an emergency or we perceive an existential threat, as in 1961 when the Apollo decision was made. That's not the case today for the government's civil space efforts.

U.S. national space policy is notable for its consistency over the past half century regardless of the party affiliation of the president. Certain basic tenets have been with us since the early days of the space age and are widely accepted as uncontroversial, such as the commitment to explore and use space for peaceful purposes, the rejection of claims of sovereignty in space, the importance of scientific discovery, and the desirability of international cooperation.

Top-level guidance on space, addressed in a reasonably comprehensive manner, originated in the Dwight Eisenhower administration in the late 1950s. Subsequent administrations over the next two decades didn't pursue broad space policy documents, preferring instead to touch on specific space-related issues in short (typically one or two-page) National Security Action Memoranda (under John F. Kennedy and Lyndon Johnson) and National Security Decision Memoranda (under Richard Nixon and Gerald Ford). The administration of Jimmy Carter marked a return to a more expansive policy with a national space policy directive signed by the president in 1978. The Carter administration also issued separate directives on civil space policy, remote sensing policy, and space nuclear power systems, all of which were signed by National Security Advisor Zbigniew Brzezinski. Since then, each occupant of the White House has redrafted and reissued a top-level space policy document. Although rewriting of the "comprehensive" space policy has become conventional practice, an obvious problem with this approach is that the space enterprise has grown and diversified so much that no single policy document can be truly comprehensive.

The Ronald Reagan administration's first national space policy was issued on July 4, 1982, the same day that the fourth space shuttle mission returned to Earth and the Space Transportation System was declared operational. Space advocates believed this signaled the dawn of a golden age of space development and commerce. But the hoped-for initiation of a space station program was absent from Reagan's policy, and future migration into space was only hinted at by one of his key civil space objectives: "continue to explore the requirements, operational concepts, and technology associated with permanent space facilities." By the time Reagan's second national space policy was issued at the beginning of 1988, the space station program was well underway and the space shuttle program was in the process of recovering from the *Challenger* accident. But the language of the new policy regarding human spaceflight was only vaguely encouraging, simply calling for the establishment of "a permanently manned presence in space." The national space policy of the George H. W. Bush administration, which appeared in November 1989, retained the same wording as the Reagan policy on this topic. The next revision came from the Bill Clinton administration in September 1996, but it was no more encouraging, and was interpreted by many as pushing the timeline further to the right by stating that the U.S. would "establish a permanent human presence in Earth orbit. The International Space Station will support *future decisions on the feasibility and desirability* of conducting further human exploration activities." [emphasis added]

George W. Bush's exploration plans were captured in a stand-alone policy issued in January 2004 rather than in his national space policy, which wasn't completed until August 2006. He took the traditional destination-driven approach, calling for a return to the Moon and then onward to Mars—the same plan that his father, the elder George Bush, had proposed during his presidency 15 years earlier. The younger Bush made only token mention of the role of international partners and the U.S. commercial sector.

John Marburger, the science advisor to Bush throughout his presidency, provided behind-the-scenes insights on the formulation and implementation of the 2004 exploration policy when he testified to the Augustine Committee in August 2009. The end result seems to have deviated from what Marburger would have preferred. Four months before the release of the exploration policy, Marburger's office provided a report to agency leaders recommending that:

> A balanced approach would prudently build capability for human exploration beyond low Earth orbit, while avoiding premature commitment to specific large scale operations. By careful planning we can *make each step a foundation for a range of next steps*, so with time our investments mount, costs and risks diminish, and we keep options open to exploit the

right one when we are ready to make a big move ... [This approach] *optimizes not for a single mission but for the steady accumulation of technologies and capabilities* that provide a base for multiple operations. [emphasis added]

The stress is rightly placed on building capabilities before picking destinations. But what did they have in mind when they wrote of keeping options open and making "a big move"?

Imaginative schemes have been proposed for further enhancing our national security, economic strength, and scientific knowledge using objects and phenomena accessible only from space. Those that require large structures beyond Earth's orbit are impractical today. *If such schemes are ever to be realized, the groundwork must be laid far in advance of their implementation. Laying that groundwork is the task of our generation.* [emphasis added]

Dr. John Marburger, the late physicist and presidential science advisor whose views on space development should have been given more weight by decision-makers.
Source: Brookhaven National Laboratory

Part of Marburger's approach was to seek the appropriate balance between human and robotic efforts. The report found that "human space flight [other than the study of the effects of weightlessness] does not contribute in a tangible way to the immediate task of preparing for large-scale development of space resources" and recognized that we do not currently "possess the knowledge or technical infrastructure necessary to deploy humans safely beyond Earth's immediate neighborhood." The step-by-step philosophy "introduces human capabilities only as appropriate, keeping in mind that the ultimate goal is to permit humans to operate routinely on missions where they are needed." This position is more forward-looking and more expansive than the oft-stated "ultimate goal" of putting people on Mars. As Marburger told the Augustine Committee, "In my opinion the all-encompassing scope of the vision we advanced was diminished in the final policy by specific mention of Mars as a target."

Lest anyone think that Marburger's approach lacked ambition, note that it embraced big, long-term projects:

> *The approach entails extensive robotic exploration of the Moon*, potentially followed by the construction of a permanent lunar base whose objective is resource exploitation possibly for economic gain and to use the material to facilitate further expansion of human exploration deeper into the solar system. This program would be carried out simultaneously with research into the mitigation of human risk factors, and the development of new launch and transport technology. [emphasis in original]

The report demonstrated an appreciation for the role of humans in space development, but didn't exaggerate that role.

> Future desirable large-scale operations in space, such as resource exploitation, climate control, and solar energy schemes, will probably exceed the capacity of robotic systems for independent operations. Under these circumstances, human participation can be justified and will probably be required.

Marburger's 2009 statement to the Augustine Committee made clear his dissatisfaction with the way the program had evolved since his office made its recommendations.

> It would be a mistake to assume that the actual development path for space exploration since 2004 has accurately reflected the overall concept of the *Vision* . . . NASA decisions following 2004 tended to diverge from the compromise *Vision* toward greater emphasis on a Mars expedition as a primary objective, and minimizing the features of sustainable, cumulative capabilities and commercial participation that were important parts of the *Vision* . . .

Taking a destination-driven path and presenting commercial and international collaboration as an afterthought in the 2004 exploration policy contributed to the policy's critical error of artificially separating space exploration from space development—two parallel activities that draw on the same knowledge base, operational experience, and technologies that will enable us to extend our reach across the solar system. Fortunately, President Obama didn't repeat this error when he came out with his own national space policy in June 2010. But regrettably, the Obama policy did not embrace the sage advice that Marburger left behind, missing an important opportunity to provide the depth of purpose and the sustainable approach that are still lacking.

Obama's revision displayed the aforementioned consistency of policy content, introducing changes that were mainly in tone and emphasis. Most notably, the new policy was more forward-leaning in international cooperation and gave higher priority to stimulating U.S. commercial space industries, conducting Earth science, and using space sustainably (that is, taking greater care to mitigate orbital debris, sharing more data on space traffic, and taking a cautious approach to development of offensive and defensive space systems).

The language on space exploration and development in Obama's policy was just over 300 words, not sufficient to guide NASA's flagship activities through the years ahead. The policy directed NASA to aim for human visits to an asteroid by 2025 and to Mars orbit by the mid-2030s. (These milestones had been mentioned earlier, in a speech by the president at the Kennedy Space Center on April 15, 2010.) Once again, a U.S. president chose to rely on the traditional destination-driven approach to define the nation's human spaceflight goals. As in previous administrations, this approach was used because it's familiar, it's easy to understand, and it seems to absolve the decision-makers of any further need to justify or expand on the vision. That includes explaining its underlying purpose and the benefits it brings to the nation and the world that are worth the cost and risk. Lacking a well-formulated rationale, it would have been wise to leave out of the policy any mention of human missions beyond the Moon until their specific purpose and their place in a long-term strategy could become more firmly established.

If we need a well-formulated rationale, who should be its architects? How would it be represented in a national policy? What are the chances that significant advances can find their way into policy in a timely manner?

The idea factory

As discussed in the previous chapter, it's important to recognize that President Kennedy's leadership example at the beginning of the Apollo program, which is interpreted (not entirely accurately) as top-down policy-making, occurred at an anomalous time. Space technology advances, and concepts for their application, typically evolve from the bottom up. New ideas originate from within the space-related agencies, or from other parts of the community such as corporations and other non-governmental organizations (NGOs) like universities, non-profit institutions, industry associations, and citizen advocacy groups. These ideas must be channeled through an internal government process to get them incorporated into national policy because NGOs do not have a seat at the decision-making table. However, they do have a variety of informal processes that allow them to contribute.

On rare occasions, NGOs and the interested public are formally asked for input. An early example of this occurred in 1985-86 as the congressionally mandated, presidentially appointed National Commission on Space contemplated civil space policy and strategy for the coming 50 years. Comments and suggestions were accepted in written form and at the Commission's town hall meetings around the country. Whatever influence that input may have had on the content of the Commission's report, it didn't drive strategic planning largely because of unfortunate timing. The report was released a few months after the *Challenger* accident, and the Reagan administration didn't take any significant implementation actions before leaving office.

A more recent example of public input comes from another presidential commission, this one a 2004 effort to plan the implementation of the Bush administration's exploration policy. The study panel, known as the Aldridge Commission, set up a website to accept ideas from individuals and organizations. In this case, the policy was already in place, so the outreach effort was designed to solicit public support and collect implementation suggestions. The 2009 Augustine Committee also had an online system for comments. The Internet has made widespread participation possible, but for study panels such as these, which exist for only a few months and have limited resources, one wonders how effectively they are able to review and incorporate the huge volume of responses.

In the old days, when NASA was young, political scientists talked about an "iron triangle" as the model for how public policy decisions got made.

Using the early U.S. space program as an example, the three points of the triangle would have been the executive agencies involved in space (NASA and the Department of Defense), the relevant congressional committees, and space interest groups (at that time, primarily large contractors). Each of these players, theoretically, acted in ways that reinforced the interests of all three, with the money flowing from the Congress to the agencies and then to the contractors who created jobs and prestige for the appropriate congressional districts. This model has compelling aspects, but it's a fairly rigid policy monopoly that couldn't be sustained for long. Policy issues, even complex, specialized, relatively new ones like space, will see their interest groups expand and diversify, breaking open the triangle and disrupting the clean interfaces between the traditional participants.

The emergence of a large, multi-faceted NGO community, whose members vary significantly in their capacity and willingness to act on key issues, demands a new, more detailed model. One possibility is the "issue network," suggested by political scientist Hugh Heclo in 1978. It consists of a large number of participants, with varying degrees of commitment, moving in and out of the network continuously, depending on their interest in particular issues. It's difficult to determine where the network begins and ends, and no one person or group appears to be in control of the policies and issues. Sure sounds messy—and a lot closer to reality.

Government decision-making on space is dominated by internal processes, supplemented by information from organizations and individuals that have established their credibility as valued information sources, such as corporations, professional associations, academic and research institutions, think tanks, international organizations, and citizen advocacy groups. There is no doubt that NGOs have an effect on public policy related to space, but it's usually slower and more subtle than many of them would wish. Claims of swift and direct influence should be treated with skepticism. For example, an advocacy group may claim that it single-handedly saved a government program from cancellation, or that it deserves credit for initiation of a new program. This is exaggeration in all but the most extraordinary cases. There is little evidence that direct interventions by NGOs, particularly citizen advocacy groups, have been necessary or sufficient to ensure positive outcomes in public policy decisions on space.

On the other hand, it's clear that space-related NGOs constitute the majority of the active space community worldwide, vastly outnumbering their government counterparts. Collectively, that makes them the keepers of the culture. They shape it and preserve it through a multitude of conferences, workshops, publications, and other formal and informal contacts across programs, disciplines, national borders, and generations. They are the primary sources of expert information and new ideas. Public policy on space would be bankrupt without them. But this role requires the patient development of networks and credibility over an extended period of time.

Space tourism is an example of an idea that needed to be cultivated for a long time in the NGO community before being accepted in public policy. Only recently have governments recognized that this is an endeavor to be nurtured, requiring their near-term attention on licensing and regulation. It took three decades of evolution in technologies, economics, and attitudes to overcome the giggle factor and make space tourism an accepted activity that governments could no longer ignore. Other examples include global satellite navigation and the Hubble Space Telescope, both of which eventually became reality as a result of 30 years of conceptual, technical, bureaucratic, and budgetary development that had to overcome perceptions that they would not be worth the investment. Direct intervention by NGOs did not produce immediate policy or programmatic actions in these areas, but patience and persistence allowed them to become mainstream activities.

So we have this big, active community of people whose ideas and visions for the future are bubbling up over periods of months, years, or decades. Once these brilliant creations are fully ripened and ready for picking, the next logical step is to present them to the large, well-funded staff office in the Executive Office of the President that is dedicated to space issues and is eagerly awaiting creative inputs that can be developed to serve the national interest. The problem is that there is no such office.

As the nation that spends far more than any other on space, and is dependent on space for its security and economic health, certainly the United States must have a focal point for all this activity in the White House. For many years, that focal point has been one position on the staff of the National Security Council (NSC). For the latter part of the Bush administration and the first year and a half of the Obama administration, that person was Peter Marquez, who tells an illustrative story from a 2008 trip to Japan. Peter went to Tokyo to meet his Japanese counterpart, who showed up with a sizable entourage that he introduced as his staff. When asked if he had brought any of his staff along, Peter answered, "You're looking at him!"

The lone NSC space official typically works closely with one or two staffers who spend at least part of their time on space issues in the Office of

Science & Technology Policy (OSTP, the office of the president's science advisor). The OMB also has people who work on space issues, but their job is formulating budgets, not policy (although they have a history of involving themselves in the latter far more than the agencies would like). The bulk of the government's space policy expertise is scattered around the agencies, and it's the job of that small cadre of White House folks to bring them all together and reach consensus on a plan that serves the nation's interests. It's not exactly like herding cats, but there are similarities. Each agency is responsible for looking out for its own programmatic, bureaucratic, or diplomatic interests, so there's a high probability of conflicts large and small, and working them out takes time.

Formulation of George W. Bush's national space policy took nearly three years. (I had a small role in that activity for about a year.) After the first complete draft of the document was created, the NSC started holding weekly meetings with all the interagency representatives around the table—20 to 30 people, picking apart the wording line by line. The meetings were considerably smaller toward the end, which came at Revision #35. All that remained after that was to get the senior officials at all the participating agencies to sign off on it, and then send it up the chain in the NSC and convince the National Security Advisor that it was ready for the president's signature. Sounds simple, but this too takes a while.

A different approach

President Obama's national space policy was initiated by a study directive in early 2009, which spawned a request from the NSC for written inputs on agency positions and ideas on an assortment of key issues. (Once again, I had a small part to play in this activity.) In the process that followed, the NSC and OSTP staffers did the writing and solicited feedback and further input from the agency representatives—but on an individual or small group basis, not with weekly meetings of a large conclave. The emergence of the new policy just a year and a half after Obama's inauguration was considered quick work.

As the Obama administration was formulating its approach to space policy during that period, I advocated to colleagues in the space community that the president should issue a stand-alone space exploration policy as the Bush administration had done. But this time, a better effort was needed to explain the purpose and long-term goals of the undertaking. In simple terms, a policy is a statement that says "here's what we're going to do, and why we're doing it." (Strategy, the next step, says "here's *how* we're going to do it.") A simple concept, but decidedly difficult to turn into a document that's coherent and substantially complete. What we need, I suggested, is a single document that integrates space exploration and development, and proposes what we want to accomplish with this activity well into the future.

I got plenty of head-nods from people in agreement that this would be a very useful thing to develop. Not surprisingly, since this would be breaking new ground, I also got plenty of blank stares when it came to talking about what could fill the pages of such a document and achieve sufficient agreement. It was clear that I needed to take a whack at it myself.

The basis of the government's role in space development should be articulated in a long-range national policy with clearly defined approaches to managing the evolution of the civil space sector and facilitating the growth of the commercial space sector. In suggesting what such a policy should look like, I have no delusions that this has a chance of being adopted wholesale. I know from personal experience that the probability of new ideas getting accepted in government policy correlates directly to how closely they resemble the old ideas. But if you want forward motion, you've got to start by injecting fresh thinking into the debate. I offer these ideas to stimulate discussion on how to create the best possible future in space. Maybe in a few years they'll bubble up to the point of being taken seriously.

For starters, we must recognize that space exploration and development will not evolve the way they did in the Cold War era. The U.S. government should not be expected—and in fact, is not able—to fund, develop, and operate all the needed research projects, services, and infrastructure. The community of participants in research, operations, and funding needs to be enlarged. Many people are recognizing this, but it still needs to be better reflected in the principles and goals that are the essence of a good policy.

Space evolution in the 21st century will require a transition away from the way we've planned and prioritized our space activities in the past, so we need to start by articulating the principles that will guide actions under our hypothetical U.S. exploration and development policy:

- Space exploration and development shall be undertaken for the purposes of increasing scientific knowledge, improving stewardship of Earth, adding value to the global economy, enhancing international cooperation, and in general, extending human activity into the solar system for peaceful, beneficial purposes.

- Government-funded space infrastructure projects shall have applicability beyond a single mission or short-term series of missions.
- New operational capabilities and infrastructure created in U.S. government space development programs shall be designed for transfer, as early as possible, to operational entities in the U.S. government, private sector, or nonprofit sector.
- Operations beyond limited-duration science missions and engineering test projects shall not be assigned to NASA or other U.S. government research and development (R&D) organizations.
- U.S. government exploration and development missions will include humans when their presence is expected to yield cost-effective benefits or otherwise uniquely contribute to mission success and/or the national interest.

These principles establish a philosophy and work environment that facilitate the concurrent evolution of exploration and development employing partnerships between government, nongovernment, and international players, each performing the roles most appropriate for them. The next step is to set ambitious but achievable goals that progressively add space capabilities and contribute to global solutions.

The new goals must be more precise than in past policies, which typically have called for broad, ill-defined actions like advancement of U.S. interests and expansion of human activity into the solar system. But precise doesn't have to mean restrictive or lacking in ambition. In fact, the opposite is true—the goals must be both flexible and bold. They should be viewed in two different timeframes: short—to medium-term (2010s to 2030s) and long-term (2040s to the end of the century). Also, they should be recognized as goals that draw together the efforts of the whole nation, since this enterprise will demand more than can be achieved solely through government programs and investment.

Short—to medium-term goals should seek to develop enduring infrastructure, skill sets, and experience that will be essential for living, working, establishing communities, and creating value in the inner solar system. These goals are capabilities—not destinations—that will be essential for creating a spacefaring society that can expand its knowledge, economy, and sustainability. They include developing the technologies, processes, expertise, and infrastructure for:

- Utilizing the unique characteristics of space, such as microgravity, vacuum, high-intensity solar exposure, and isolation from Earth, to produce useful knowledge and products.
- Harvesting and processing of extraterrestrial materials and energy resources.
- Building progressively more sophisticated structures in Earth and lunar orbits.
- Building installations on the Moon, constructed to the greatest extent possible with local materials.
- Advancing space robotics to minimize the need for human presence in activities that are hazardous, remote, or are strong candidates for automation, and to provide direct assistance to humans where human involvement is required.

Achievement of these goals should lead to the following long-term goals starting around mid-century:

- Construction and operation of advanced structures that minimize their dependence on supply lines from Earth, designed for science, commerce, and other purposes.
- Aggregation of space structures into industrial parks at locations deemed valuable for their proximity to space resources, Lagrange points, or other attributes.
- Realization of significant contributions to the terrestrial economy through energy and manufactured products for use on Earth and in space.

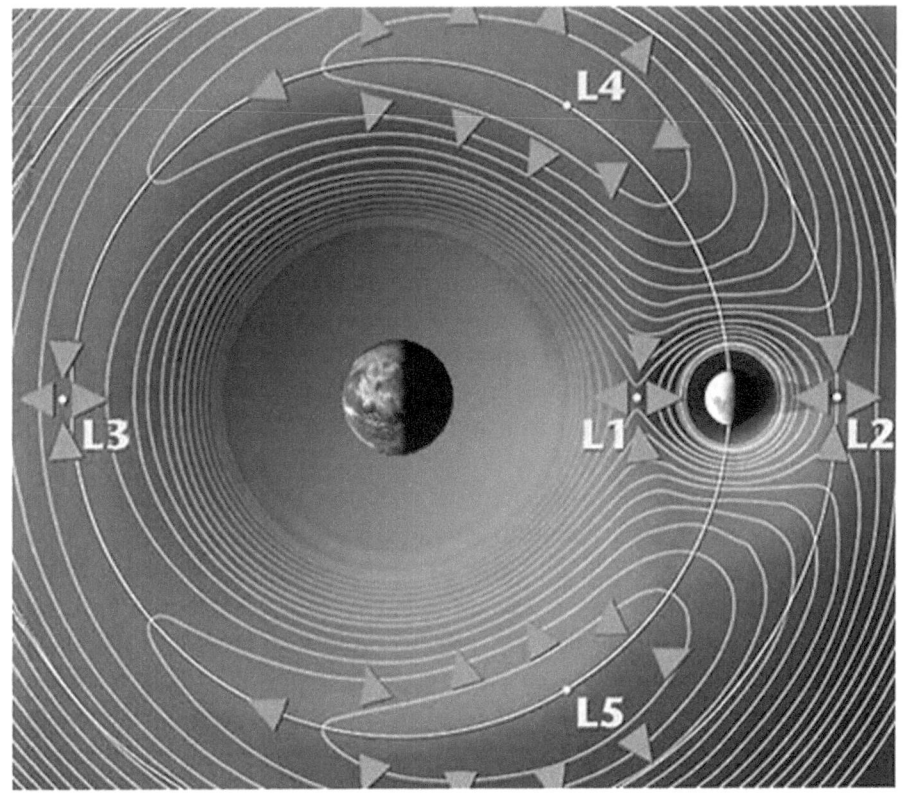

Lagrange points in the Earth–Moon system. NASA is considering missions to L1 and L2 in the 2020s.
Source: NASA

Note that none of these goals specifies a planetary destination beyond cislunar space. Certainly, the Moon and near-Earth objects will be early destinations due to their close proximity and the broad range of contributions they can make to the goals. What comes next, and when it should come, should be driven by progress toward the goals, the rate of technological advance, the lessons of experience, and the availability of resources from all participants.

If the principles and goals, as suggested above, require that ongoing operations be kept separate from research organizations, and that our top priorities are advanced knowledge and useful capabilities rather than specific destinations, we can begin to piece together a strategy that will put us on a path to our goals. At a minimum, such a strategy would require the following:

- Redirect NASA to focus more exclusively on R&D and shed its operational duties, other than limited-duration science and engineering mission operations that support its R&D programs.
- For operational and infrastructure components created by U.S. government space development programs, establish the identity and relationship of the intended system operator at the beginning of the program. The intended operator may be an operational entity of the U.S. government, the U.S. private sector, the U.S. nonprofit sector or an international consortium, but not an R&D organization.
- Set target dates for achieving capability milestones. Plans for reaching planetary destinations shall flow from the achievement of relevant capability milestones, not the other way around.

Principles and goals should be designed to endure, while the strategies and programs supporting them should be allowed to evolve. A document providing top-level policy guidance, such as a presidential directive or authorizing legislation, should not get into specific programmatic decisions, which must be allowed sufficient flexibility.

The political and organizational difficulties of executing a significantly altered U.S. civil space policy should not be underestimated, as demonstrated by the political clashes of recent years, described in previous chapters. (And those battles were prompted by the Obama administration's proposals for NASA in the 2011 budget request—changes far more modest than those I've just suggested!) Major components of NASA would be profoundly changed. The agency has become accustomed to designing and procuring its own space systems and operating its own launch facilities. This is a natural result of the fact that for most of the agency's history, there was no other entity willing and able to do all of this for them, especially in the case of human spaceflight. That is changing in the 21st century, and ultimately NASA must embrace that which it helped to create. If these new space services can do the job, this will mark a new level of spacefaring maturity spawned by NASA's stimulation of technologies and markets. This is one of the big payoffs we've been waiting for, so NASA should feel proud rather than threatened.

Does this proposal go too far, given the short planning horizon that's typical of our political system? It may seem so, but there is precedent for taking a longer view. For example, when the U.S. Navy plans for the future of its fleet, especially complex big-ticket requirements like an aircraft carrier group, it must think through the entire life cycle—at least 30 to 40 years—including development, deployment, decades of operation, and retirement. That means today's Navy planners are working in a timeframe that stretches through mid-century. Should we expect any less from the planners of our space infrastructure?

Chapter 10

The Next Great Thing

If the architecture of the exploration phase is not crafted with sustainability in mind, we will look back on a century or more of huge expenditures with nothing more to show for them than a litter of ritual monuments scattered across the planets and their moons.—Dr. John Marburger, Science Advisor to President George W. Bush, 2008

On November 8, 2011, the Earth experienced the close flyby of a sizable asteroid with the unmemorable name 2005 YU55. In this case, "close" means about 320,000 kilometers, less than the distance to the Moon, and "sizable" means 400 meters across, big enough to do plenty of damage if it smacked into a populated area of our unprotected planet. Fortunately, we knew for years that it was coming, and that it would miss us. This allowed time for the science community to plan a series of observations using ground-based optical and radar instruments, including NASA's Deep Space Network and the big Arecibo radar dish in Puerto Rico. (The Hubble Space Telescope was not used because it wasn't designed to track objects moving quickly in its nearby field of view.)

The fortuitous arrival of this celestial visitor provided a target of opportunity for astronomers and planetary scientists, but it also pointed out a shortfall in our capabilities despite a half-century of spaceflight. Even though we had years of advance notice, we had no way of sending a spacecraft to meet 2005 YU55 for observation, sample retrieval, or if necessary, deflection. Admittedly, it would have been technically challenging to match the trajectory and velocity of the asteroid, and politically challenging to find a way to pay for the mission. But the same is true for asteroid intercept missions that we attempt at interplanetary distances. Unfortunately, we're still at a primitive stage in which we can't pull a robotic spacecraft off the shelf and launch it quickly, or hop into a real-life counterpart of the *Millennium Falcon*, and shoot over to an asteroid that's passing through our neighborhood.

Come to think of it, we can't even do things that should be much more routine, like repair and refuel our expensive satellites. Or retrieve derelict satellites and used rocket bodies so they don't linger as debris hazards, or worse yet, break up into thousands of debris hazards.

The previous chapter introduced a depiction of three stages of space development: training ground, industrial park, and settlement. Obviously, we're still in Stage One, seemingly a long way from reaching Stage Two. How long it takes us to advance to the next stage is dependent on good choices, sustained commitment, and a reasonable amount of good luck. At this point in our development, it's critical that we make the right strategic choices as we decide what will be the Next Great Thing in space. By now it should be no surprise that I don't believe manned missions to destinations beyond the Moon are the logical next step. At this stage, they would be a resource-sucking distraction. Nor should we squander scarce resources on big hardware projects that have been created in isolation from a well-conceived long-term strategy. The Next Great Thing in space should be to move smartly toward Stage Two, turning cislunar space into an active industrial park.

Though less dramatic than headline-grabbing milestones like "the first humans on Planet X," striving for Stage Two will challenge us to build new and better space capabilities, expand the frontiers of science, bring direct benefits to Earth, and eventually enable us to achieve Stage Three. The secondary benefits, including technology spinoffs and inspiration of our youth, are likely to be realized in great abundance, beyond anything we achieved in the aftermath of Apollo. But the enterprise must be justified by its primary benefits, and the gifted minds of post-Apollo generations seem to have their own perspective on what those benefits might be.

Visits to exotic locales are still appealing—but for vacations, not as career goals. Be the first to reach Earth's poles or the summit of Mount Everest? Ancient history. Fly above the clouds at high speed? Millions of people do it every day. Put the first footprints on the Moon? Old news. How about creating advanced technologies for remote sensing, orbital debris cleanup, energy distribution, and extraterrestrial resource mining? Hey, now you're talking. Let's do things that raise living standards, improve our stewardship of the Earth, and lead to purposeful expansion in the solar system. We can increase the capacity and broaden the worldview (universe-view?) of both the individual and the human collective. It's useful, it's cool, and it might even make money.

The path to Stage Two

So how do we get to Stage Two from here? We already have an assortment of Earth-to-orbit launchers and satellite fleets that perform a variety of valuable functions for our security, our economy, and our scientific pursuits. All of this has made an immense contribution to improving the human condition, but it's just the beginning. The next step will require more than evolutionary improvement of the things we're already doing, which mostly consist of transmitting electrons back and forth. We'll need to be able to harvest and process raw materials and energy in space. We'll need to build things, large and small, in orbital space and on the Moon. We'll need laboratories, manufacturing facilities, and habitats. We'll need the means to efficiently transport people, cargo, and automated systems throughout cislunar space. Vital to all of this is proximity operations (often shortened to "proximity ops"), the ability to rendezvous, dock, and otherwise conduct activities involving separate spacecraft interacting in close proximity or direct physical contact.

Identifying proximity ops as the key to our future in space may seem odd since we've been doing rendezvous and docking of spacecraft since the 1960s, and it's a routine part of space station operations today. However, the bulk of this activity has been associated with high-profile human missions in LEO, and in lunar orbit during the Apollo era. It's the spaceflight equivalent of brain surgery—it's done by a small cadre of experts, and very, very carefully.

There are no operational manned or robotic systems that can perform rendezvous, capture, repair, refueling, reboost, and retrieval of orbiting payloads throughout cislunar space. Since the retirement of the space shuttle, there isn't even an operational system that can do these things in low Earth orbit. Such systems are needed if space development is to advance beyond its current stage. This will be true across the civil, commercial, and national security space sectors. NASA researchers at the Glenn Research Center in

Cleveland have suggested, for example, that a large-scale "Space Harbor" for satellite servicing is "an essential economic pre-condition and next parallel or sequential step on the road toward exploration beyond LEO."

The idea to develop space systems for proximity ops is not new, at least for low-to-medium altitude Earth orbits. NASA studied such systems in the late 1970s, when it became evident that the space shuttle would not be flying in time to rescue the Skylab space station, which was falling out of orbit and could no longer count on Apollo flight hardware for a reboost. But Skylab came down before NASA had the time or the funding to develop a robotic rescue vehicle.

The idea resurfaced in the mid-1980s as a way to extend the reach of the space shuttle. NASA awarded a contract to TRW Corporation in 1986 to build an Orbital Maneuvering Vehicle (OMV), which was intended to ride into orbit in the shuttle's cargo bay and then use its own power to rendezvous and dock with a variety of spacecraft, such as the Hubble Space Telescope, to provide reboost, retrieval, and remote viewing and servicing using tele-operation from the space station or from the ground. It was to have a 10-year design life and was intended to reach altitudes up to 1,500 miles, change its orbital plane (angle relative to the equator) by up to eight degrees, and hibernate on orbit for up to nine months. It was envisioned with special-purpose mission kits that would allow satellite refueling, change-out of batteries and other components, and interception of orbital debris. Also on the drawing board was a module designed to house a space-suited astronaut.

Within three years, the estimated cost of the OMV had grown more than 80 percent, so in December 1989 NASA scaled back the project's goals. The OMV's on-orbit stay-time and maneuvering flexibility were reduced, and it would no longer perform avionics servicing or carry an inspection camera. However, it still would have retained the ability to boost the Hubble and other large spacecraft.

A few months after NASA curtailed the program's ambitions, the General Accounting Office (GAO, now called the Government Accountability Office) issued a report which found that cost increases were likely to continue despite the reduction in capabilities. Additionally, the GAO stated that "A firm requirement for the OMV does not exist." Planned missions such as reboosting Hubble could be accomplished "in other ways that cost less." The GAO pointed to NASA studies showing that the space shuttle could maintain the Hubble and other planned satellites "at acceptable altitudes by reboosting them during regularly scheduled maintenance flights. One or two additional shuttle flights dedicated to reboosting the observatories could be required early in the next century if the OMV is not developed." In its draft report, the GAO recommended cancelling the OMV. Six days after receiving the draft, the NASA administrator terminated the program.

The GAO estimated that completion of the OMV, with all of its original capabilities restored, plus the ability to launch on vehicles other than the space shuttle, would have a total cost of $1.3 billion. In hindsight, that seems like a bargain—we shouldn't have passed up this important first step toward Stage Two development. More than two decades later, we still have nothing like it.

Technical and funding challenges aren't the only reasons we haven't progressed as much as we could have in this area. Proximity ops are a touchy subject in many parts of the space community. Despite their potential benefits, commercial operators and insurers worry about the risk of collision or other "oops" moments that could damage a spacecraft's components or send it spinning. This could permanently end a revenue stream of millions of dollars per month for a commercial satellite and cause a debris hazard that the operator would be responsible for.

National security operators share the same concerns about the risk of service interruption or permanent damage. In addition, they'd rather not see a proliferation of capabilities for getting up close and personal with their satellites. In fact, they'd prefer that no one else knew exactly where those satellites were, or even the fact that they existed.

Anti-satellite weapons (ASATs), which can be seen as a specialized application of proximity ops, have prompted concerns since the beginning of the space age. The more we improve guidance, navigation, and control systems—and the more those systems proliferate—the greater the concern becomes. This already was evident in the early post-Apollo years, when the Soviet Union was testing a co-orbital explosive ASAT. In response, the administration of Jimmy Carter opened negotiations on ASAT arms control, during which the Soviets raised objections to the forthcoming space shuttle, which they labeled an ASAT weapon system. More recently, experiments in autonomous proximity ops by the Air Force (XSS-11), NASA (Demonstration of Autonomous Rendezvous Technology, or DART), and the Defense Advanced Research Projects Agency (Orbital Express) were interpreted by some observers to have at least secondary objectives in ASAT development.

It seems abundantly clear that if we have any plans of advancing the development of space, we'll need to overcome our fear of proximity ops.

To prohibit or excessively restrict these activities just because they pose a risk of accidents or could function as weapons would mean halting further space development. The global community must accept proximity ops and establish behavioral norms that dispel fears and tensions. This is how we deal with other technologies such as aircraft, ships, ground vehicles, and cell phones. All of these have been used in crime, terrorism, and war, but it would be absurdly counterproductive to ban them.

International norms for proximity ops in space could be established in a manner similar to the way we've developed norms for security, economic interaction, and environmental stewardship. We've done this already for space in the form of international treaties, principles, and guidelines, including orbital debris mitigation guidelines. The negotiating process takes many years, and it doesn't always progress smoothly, but the cumulative result has been an environment that facilitates maximum interaction between space players while minimizing risks.

Under Article VI of the Outer Space Treaty of 1967, signatories (including the U.S. and all other major spacefaring nations) must "bear international responsibility for national activities in outer space ... whether such activities are carried on by governmental agencies or by non-governmental entities." Furthermore, "The activities of non-governmental entities in outer space ... shall require authorization and continuing supervision by the appropriate State Party to the Treaty." The U.S. has put in place regulatory procedures to address launch, reentry, electromagnetic spectrum use, orbital slot assignments in GEO, and debris mitigation, but not on-orbit actions such as proximity ops or debris removal. Development of cislunar space will require that behavioral norms be established for these actions as well. In the U.S., this will most likely start with national guidelines for proximity ops that will be reflected in licenses issued by the government to organizations involved in such operations. As demonstrated by U.S. transportation history, successful implementation of this process could promote growth in commercial space activities by helping to reduce the debris threat, improving safety, and easing international tensions regarding behavior in space.

The specific U.S. guidelines that would emerge from extensive interagency discussion and debate may include, at a minimum:

- Prohibition against interference with satellites that have not been offered up for salvage. (A proposal for a new approach to space salvage is discussed below.)

- Prior public notification of launch or orbital maneuvers to initiate satellite servicing and retrieval missions.
- Prior notification to satellite owners regarding proximity operations to be conducted within a specified distance (such as a few kilometers) of their space assets.
- Immediate alert of any servicing or retrieval mission that does not go as planned and may create a hazard for others.

Near-term plans for on-orbit satellite servicing experiments by government agencies and private sector organizations have already raised eyebrows internationally. A variety of national and non-governmental entities are planning orbital research platforms and robotic satellite servicing for later in this decade, so it's not too early to start talking about the international establishment of behavioral norms.

Sanitation engineering in space

U.S. orbital debris mitigation guidelines were developed in the late 1990s in a collaborative effort between the Department of Defense and NASA, and adopted by the National Security Council as national guidelines in December 2000. Immediately thereafter, the U.S. began the long process of gaining international acceptance of the guidelines to encourage existing and emerging spacefaring nations to use best practices that would help control the growing debris problem. This effort was eventually successful in establishing voluntary international guidelines very similar to those followed by the United States.

The calls for action have increased in recent years in the wake of several debris-creating incidents, most prominently a January 2007 Chinese ASAT test, a February 2008 U.S. intercept of a disabled national security satellite, and the February 2009 collision of an active Iridium communications satellite and a defunct Russian Cosmos.

Global adoption of best practices for mitigation is ongoing, but even broad success in this area would not provide a full solution to the debris problem. The next step, removal of debris, has been discussed for decades without advancing to the implementation stage due to technical and affordability limitations. Policy and international legal concerns have been identified, but these remained in the background as the formidable technical challenges pushed the testing and deployment of remediation systems well into the future.

Operational debris removal systems may no longer be such a distant prospect. New proposals for system designs are appearing frequently. Advances in robotics, satellite bus design, automated rendezvous and docking, and low-mass orbital maneuvering systems, coupled with a variety of efforts to reduce launch costs, may make debris remediation practical in a few years. If realistic technological solutions are starting to appear on the horizon, it's time to direct our attention to the two most significant hurdles in policy and international law that need to be surmounted if remediation efforts are to be successful: 1) international law that treats salvage in space differently from salvage at sea, and 2) remediation technologies and operations that look like and could double as ASAT systems.

Given the degree of importance assigned to the debris problem today, it may seem surprising that there were no meaningful, successful actions to promote good practices on the international scene throughout the Cold War, even in the period from the mid-1960s to the mid-1980s when numerous space treaties and principles were enacted. While debris was a concern, it was not seen as an imminent threat requiring broad actions by the major players. Furthermore, the U.S. and the Soviet Union were not inclined to seek compromises that involved sharing sensitive information about space system operations and plans for orbital tests with the potential to cause debris.

The Outer Space Treaty established the Cold War's only rules governing the treatment of orbital debris. Article IX, which is primarily concerned with contamination from extraterrestrial matter, is generally interpreted to be applicable to orbital debris as well, due to language that directs "appropriate international consultations" prior to engaging in activities that could cause "potentially harmful interference with activities of other States Parties." To address the sensitivities of the two superpowers—each worried that the other would try to abscond with its top secret satellites—the Outer Space Treaty granted perpetual ownership of space objects to their launching state, even after the objects are deactivated and become uncontrolled junk. Although this is an obstacle to effective cleanup efforts, most active spacefaring nations (including the U.S.) are reluctant to suggest changes to the treaty despite the fact that amendments can be offered by any signatory.

Article VIII of the treaty specifies that ownership of a space object stays with the launching state no matter where it's found or whether it's brought back to Earth. Any signatory parties, including non-governmental entities subject to such parties, that want to salvage space objects they don't own must do so with the permission of the owner. This limitation is more strict than salvage at sea. Eventually, as space operations become more sophisticated and active removal becomes a practical way to address the debris problem, the space salvage restriction will need to be addressed in some manner to allow more flexibility. When this constraint was negotiated in the 1960s, diplomats were not thinking about establishing a business-friendly environment for space salvage, and diplomats today will not do so unless the required technologies, a plausible business case, and political feasibility are within sight—or they perceive the need to address a serious threat.

For a long time, many analysts believed that small debris should be the primary objective for cleanup because it exists in very large numbers, it's difficult or impossible to track and avoid, and it's capable of doing considerable damage. But cleaning up the small stuff was a challenge with no feasible technical solutions on the horizon. Meanwhile, dead satellites and rocket bodies were seen as presenting a lesser threat because they could be tracked and avoided, so retrieval was a lower priority. This view was changing even before the 2009 Iridium-Cosmos collision as the population of derelict spacecraft and the likelihood of so-called orbital "conjunctions" increased. With the development of retrieval capabilities, the new conventional wisdom acknowledges that non-functional satellites and rocket bodies can be tracked, intercepted, grappled, and removed from orbit before they are impacted and become thousands of pieces of untrackable debris. As retrieval becomes feasible, it may be preferred over the practice of routinely maneuvering satellites out of the way of debris in an environment of increasing traffic.

Recent research at NASA, presented at international space forums, finds that the debris population would continue to grow *even in the absence of future launches* due to collisions, particularly in LEO. The only way to stabilize the population, according to the NASA researchers, is through a combination of strong adherence to existing mitigation guidelines (which has not yet been achieved) and active removal of at least five objects per year that have relatively large mass and high probability of collision. To address the removal requirement, modeling techniques have been used to create a prioritized list of hundreds of objects among the existing population of inactive satellites and spent rocket bodies. Researchers in spacefaring nations have proposed various means for interception and deorbiting of large objects, often involving techniques such as robotic attachment of tethers or small rocket engines. Ideally, a single automated spacecraft would be capable of multiple encounters, bringing down several expired satellites on a single mission.

Government or commercial entities contemplating retrieval operations must be able to choose their objectives well in advance because the missions will require detailed planning and considerable up-front investment. If the preparations involve seeking permission on a case-by-case basis from foreign governments, without the benefit of established procedures, it will be an expensive and time-consuming process that is likely to limit the available objects and undermine the already fragile economics of this activity. If the parties to the Outer Space Treaty continue to object to any attempts to update its language, then no remedy will be available in the treaty's amendment process to accommodate a modern approach to salvage in space.

Fortunately, a remedy may be available under another space treaty, the Registration Convention of 1975. Article IV of the Convention requires signatories to provide a basic set of information to a U.N. registry soon after the launch of a space object. It also requires notification when an object is no longer in space, having been deorbited or otherwise removed. There is no requirement to report anything about the object during the time between its placement in space and its removal. But although it's not required, signatories may provide input during the on-orbit life of a space object. Article IV states, in part, that "Each State of registry may, from time to time, provide the Secretary-General of the United Nations with additional information concerning a space object carried on its registry." The nature of the "additional information" is not specified in the Convention, but it could include notification that an object, though still in orbit, is no longer functioning and is not expected to be reactivated. Another possibility is that an active satellite could change ownership through a commercial or intergovernmental transaction, transferring the responsibility for that satellite to another nation.

The Convention's signatories already agree that orbital debris is a problem in need of attention. If they can take the next step and agree that it's time for action to enable debris cleanup, they could create a separate category in the registry for expired satellites and rocket bodies, labeling them "available for salvage." To date, expended hardware has been allowed to remain in orbit for many years, posing a collision hazard and fragmentation risk. As remediation techniques become available, signatories could be encouraged to put their space objects on the "available for salvage" list as they expire. In doing so, they would signal that "if you haul it away, it's yours" but would retain ownership responsibilities until a successful retrieval mission was performed. If an object is salvaged, then the original owner is relieved of responsibility (and potential liability) for that object; if no retrieval is attempted, then the current responsibilities and mitigation guidelines would continue to apply. More detailed considerations would need to be worked out as this process is established: At what point is ownership and liability transferred to the salvager (e.g., first contact in orbit or completion of the retrieval mission)? If the salvager is a non-governmental entity, how is treaty responsibility transferred to the salvager's country? Is this accomplished by prior arrangement between countries, or specified in a license and/or contract if a private entity is involved?

The salvage list should be open to all interested parties. Governments and commercial entities willing and able to attempt retrievals should be encouraged to report in advance any intended retrievals to avoid conflicts between pursuers of the same object. Salvage targets should not be "reserved" for a particular operator—at least, not until a retrieval mission is underway—because this could lead to a situation similar to the "paper satellites" problem at the International Telecommunication Union, in which radio frequency reservations are granted for communications satellites that will never be launched.

Launching states would be under no obligation to put their satellites on the salvage list. Sensitive national security assets, or satellites that the launching state intends to retrieve or service itself, would retain the traditional space object ownership status. However, launching states that own objects on the high-priority retrieval list (those with high mass and high probability of collision) should be encouraged in international forums such as the U.N. Committee on the Peaceful Uses of Outer Space to make them available for salvage. The United States and the European Space Agency would be key players, but it's also critically important to have Russia involved. Russia accounts for the majority of the derelict space objects on the high-priority list, so attempting to undertake active cleanup without that country's cooperation would seriously undermine the effectiveness of the effort. Other increasingly active nations such as China must be involved as well.

The physics and technologies involved in debris cleanup will never allow it to be completely divorced from any connection to ASATs. However, if the salvage list procedure described above is employed, the practice of satellite retrieval can be less controversial, become accepted as the norm, and perhaps stimulate a market for used satellites as debris remediation is accompanied by repair and refueling services.

A used satellite market could be a boon to small and emerging spacefarers. Refurbished satellites may be available for a fraction of the price of new ones, and pre-owned spacecraft serviced in orbit may be readied for reuse quickly. A benefit for all operators, especially those with large constellations in crowded orbits (such as communications service companies Iridium and Globalstar) would be the ability to contract with a commercial service to retrieve expired satellites if they can't be repaired or refueled, thereby eliminating the potential liability associated with their long-term presence in orbit. For large GEO constellations (such as those operated by Intelsat, Eutelsat, Inmarsat, and SES Satellite Services), the availability of on-orbit servicing, including boosting spacecraft to disposal orbits, could prevent the sacrifice of vast amounts of revenue. Currently, these satellites need to reserve enough fuel to move to higher disposal orbits at the end of their service life. If that move could be assisted by a separate robotic servicer, then the satellite's fuel could continue to be used for station-keeping, thus extending the service life—and the associated revenue stream—by many months, perhaps more than a year.

Potential objectors to salvage schemes and related proximity operations will need to be convinced that the benefits of debris cleanup—and all the other capabilities that the same technologies bring—outweigh the risks. As more nations become spacefarers and orbital traffic increases, emerging players will not tolerate it if the established operators try to limit their access to space because the orbits are too full. Rather, the space lanes will need to be tended by a conscientious global community in a coordinated effort to keep them safe for operations, in the best interests of everyone. Active removal of derelict spacecraft and other debris will have to be part of that effort in the not-too-distant future. Responsibility for coordination of the effort may reside with existing international organizations, but also could be managed by an international business collective like the Satellite Data Association, a collaboration of communications satellite operators from around the world which has proven that critical operational issues affecting both government and non-government sectors can be addressed through cooperation among competitor-colleagues.

This chapter has identified proximity operations as the key to advancing to Stage Two. It also discussed ways of improving the way we function in space, particularly by cleaning up the mess we've made in Earth orbit. But there's much more to consider. The point of space exploration and development is to extend the reach and the capabilities of traditional space applications, and to create new ones. That means conducting proof-of-concept missions, some of which may lead nowhere, some of which may be productive far beyond what we can imagine. But how do we decide which concepts we want to prove? How do we design and prioritize the missions if we haven't settled on which questions we're trying to answer?

Chapter 11

Answering the big questions

If you want a wise answer, ask a reasonable question. —Johann Wolfgang von Goethe, German writer and scientist

Space science programs have a history of being highly productive. A very important part of the recipe for success, as discussed briefly in Chapter 8, is the science community's ability to stay focused on the big, important questions that challenge their various scientific disciplines, such as: *How do galaxies, stars, and planets form and evolve? What is the history and fate of our solar system? What can we learn from studying the geology and the atmospheres of other planetary bodies? Are there Earth-like planets around other stars and do they support life?* The movement to develop and settle space, in contrast, has been far less successful in agreeing on the set of big questions it wants to pursue.

Clearly, some big questions have been articulated, though not all of them have been aggressively pursued. *What are the effects of long-duration spaceflight on the human body and mind? How can in-situ resources be used to sustain human presence on another planet?* Finding answers to questions like these tends to be harder and more expensive than carrying out pure science missions. The vast majority of the space science done so far, other than examinations of meteorites and Apollo lunar samples, has employed advanced instrumentation that collects and analyzes parts of the electromagnetic spectrum. This allows scientists to do their research without personally conducting interplanetary (or interstellar or intergalactic) field work. The big development and settlement questions can't be answered in the same way. People, and the equipment they need to live and work, must be sent into space and allowed to function there for extended periods of time.

The exploration and development community is larger and more diverse than the space science research community. In addition to scientists from various disciplines, it includes engineers, political decision-makers, businesses, investors, specialized media, and the interested public. This community needs to devise its own approach to answering its big questions in a coordinated way, akin to the way the scientific community does. Before doing that, they need to take a step back from the decadal survey approach (which prioritizes the missions that seek answers to known questions) and come to some agreement on what their big questions are. Perhaps they can be boiled down to the following:

- Can humans "live off the land" in space? If so, what integrated set of technical systems would be required to accomplish this? If not, how much can be accomplished with robotic systems alone?
- Can expansive space operations consistently create value—scientific, economic, and societal—sufficient to justify the cost and risk? What are the priorities across the various investment options?

Road maps to the solar system

As part of the Obama administration's push to reinvigorate the sagging space technology base, the Office of the Chief Technologist at NASA Headquarters created a set of 14 space technology roadmaps in 2010. Among the topics addressed were launch and propulsion systems (Earth-to-orbit and in-space), life support, robotics, nanotechnology, communications, navigation, thermal systems, landing systems, and materials. This was a welcome exercise, long overdue. The following year, the National Research Council (NRC) of the National Academy of Sciences did a thorough review of the roadmaps and issued its final report in early 2012. One of the NRC's tasks was to prioritize the large number of technology development projects that would be vying for scarce funding. In doing so, they wisely emphasized something that should be obvious: "Balance is needed between support of focused technological approaches and support of technologies that accommodate a wide range of destination and schedule options." Stated another way, mission-oriented objectives and milestones need to be served, but flexibility is also essential to promote applicability to multiple projects. We shouldn't put all of our technology investments into one-trick-ponies that are discarded before their time like the Apollo systems.

> **Top Space Technology Priorities**
> National Research Council, February 2012
> Human Sustainment Systems
>   Radiation mitigation for human spaceflight
>   Long-duration crew health
>   Environmental control and life support systems
> Spacecraft, Power, and Propulsion Systems
>   Guidance, navigation, and control
>   Nuclear thermal propulsion
>   Electric propulsion
>   Fission power generation
>   Solar power generation (photovoltaic and thermal)
>   Active thermal control of cryogenic systems
>   Entry, descent, and landing thermal protection systems
> In-Situ and Remote Exploration Systems
>   In-situ instruments and sensors
>   Optical remote sensing systems
>   High-contrast imaging and spectroscopy technologies
>   Detectors and focal planes
>   Extreme terrain mobility
> Multiple Applications
>   Lightweight and multifunctional materials and structures
>
> (The categorizations used above are added for clarity. The NRC report presented these 16 technology priorities in the framework of three overlapping objectives: sustaining humans in space, in-situ investigations, and remote investigations.)

A key concern expressed in the NRC review was that the tech roadmaps didn't adequately address the role of commercial and international partnerships. These collaborations will be an essential source of technical expertise and a critical factor in the pursuit of affordability and sustainability.

As important as it is to do tech roadmapping, it's only part of what's needed to lay the groundwork for future exploration and development. The tech roadmap tells us how we should fill our toolbox, but not necessarily what we're going to build with those tools. You could obtain the best power drill ever devised, but that doesn't tell you where to drill holes, or what useful function will be served by drilling holes. For spaceflight projects, we've traditionally relied on destination-centric answers in response to questions of purpose. Using that approach, priorities are set and designs are locked in based on the narrow goal of getting from here to there and back.

Another roadmapping effort has taken the next step, attempting to give purpose to our space toolbox and characterize the partnerships that will be needed. The International Space Exploration Coordination Group released its first iteration of a Global Exploration Roadmap in September 2011. The group is a coalition that includes NASA, the European Space Agency, and the civil space agencies of Canada, France, Germany, India, Italy, Japan, Russia, South Korea, Ukraine, and the United Kingdom. In this roadmap, we find some very hopeful signs that the collective wisdom is moving in the right direction—but it's still a work in progress.

The Global Exploration Roadmap lists several goals, including some that would be expected in any such document: perform space, Earth, and applied science investigations, including the search for life; study human physiology and expand the presence of humans beyond low Earth orbit; and engage the public in the exploration adventure. Its less common but still critically important goals include: develop capabilities and infrastructure for off-Earth operations; support commercial entities as they create new markets that bring benefits to all humankind; and pursue planetary defense and orbital debris management.

If the Coordination Group's members have the intent and find the resources to implement their plan, then it's a breakthrough in some respects. An international body representing most of the world's major spacefaring nations is endorsing more than just a continuation of prestigious human spaceflights and deep-space robotic science missions. They're recognizing that exploration and development go hand-in-hand; that robust, versatile, and sustainable space infrastructure must be built; and that benefits to Earth, through new markets and solutions to global problems, must be produced. Everyone needs to contribute brainpower, work, and investment. This requires government risk-sharing and other incentives to bring in the private sector alongside the individual and collective efforts of nations.

Despite all the refreshing talk about capabilities-driven planning, however, the Global Exploration Roadmap still has the trappings of a destination-driven strategy. Throughout the report, all roads lead to Mars, which is referred to as our "ultimate" and "driving" goal. The two alternative paths presented are "To Mars with Deep-Space Asteroid Missions as the Next Step" (Asteroid-Next) and "To Mars with the Moon as the Next Step" (Moon-Next). At least there's no Mars-Next alternative because the group sensibly acknowledged that we've got a lot to learn and to build before we get to that stage.

What's missing is an alternative that could be called "Develop Cislunar Space to Bring Benefits to Earth and Prepare Us for Our Leap Outward to the

Rest of the Solar System." That title is way too long, so let's call it Cislunar-Next. This would de-emphasize sending humans to destinations beyond the Moon for now, and make in-space capabilities, infrastructure, and experience the top priorities. Some may perceive this as a "go slow" approach, but that would be a mischaracterization. In fact, if done properly, it would be the fast track to a purposeful, sustainable future in space. It would be the push to move us from Stage One (training ground) to Stage Two (industrial park). When we reach Stage Two, the commercial sector will be on board as indispensable partners. They will have skin in the game, and their partnership—perhaps even their leadership—will propel the movement beyond cislunar space. If we rush to send humans to Mars before achieving Stage Two, the commercial sector will participate only as government contractors, and will not be there as a sustaining force.

There's a tendency to envision a sequential path that starts with observation and study, then proceeds to exploration, and finally to utilization. That's not always the way it works on Earth (for example, in the oceans), nor will it always be done that way in space. In Cislunar-Next, observation, exploration, and utilization will be conducted simultaneously.

Science fans may be concerned that this path will shortchange pure research. Not to worry—there's plenty of science yet to be done on the Moon, and deep-space robotic missions would still be elements of this approach, making it a more scientifically rich scenario than an aggressive (and resource-draining) humans-to-Mars approach. Mars would still be visited routinely by robots, and an assortment of smaller bodies would be examined close up. In fact, this scenario would compel us to address the shortcoming identified at the beginning of Chapter 10 by devising a rapid-response ability to intercept near-Earth objects. If a target of opportunity is spotted on a flight trajectory that will bring it near (or within) cislunar space, an alert force of scientists and engineers could endeavor to quickly assemble a mission that would be sent out to meet and observe it. Once it's been demonstrated that this can be done, more ambitious intercept missions could attempt landing and sample return. In addition to the science value, these would be the first steps toward creating planetary defense systems that someday may be needed to divert or destroy an incoming asteroid. (In 2012, NASA estimated there is a population of approximately 4,700 "potentially hazardous asteroids" orbiting near the Earth, of which only 20 to 30 percent have been found.)

Aside from the obvious concern we have regarding giant rocks clobbering the Earth, there are other global issues for which space can provide at least partial solutions. Some are apparent already, such as environmental stewardship, disaster warning and recovery, and scarcity of material and energy resources. Other issues amenable to space-related solutions will become evident as we build the space capabilities to address them. This Earth-focused aspect of our space future must always be prominent in our planning. The Apollo program was justified not by the science, but rather by some very down-to-Earth geopolitical concerns. Terrestrial interests will have to be factored into future space projects as well.

With all of this in mind, let's consider what the work program would include if we decided to pursue Cislunar-Next. What capabilities do we need to demonstrate in the Earth-Moon system that would be valuable elements of a space infrastructure? What are the high-priority proof-of-concept projects?

Setting priorities

Some research and development areas, such as the NRC technology priorities cited above, will be essential regardless of which approach is chosen for human spaceflight. For example, in the category of human sustainment systems, we'll need to keep improving our knowledge and techniques for dealing with the physiological and psychological stresses of long-duration missions. Life support systems need to become more reliable, lower maintenance, and less dependent on frequent resupply. This is an area that's already getting a good deal of attention aboard the ISS and in some Earth-bound studies. But we lack the facilities to do all that is needed.

The crew conditions being studied on ISS are confined to six-month stays in zero-gravity in low Earth orbit, giving us a very limited sample of what we'll encounter as we move outward. This is far short of the time needed for any interplanetary journeys, and provides minimal ability to prepare us for the radiation levels that confront us when we go beyond the shielded environment of low Earth orbit. This exposure may be the greatest potential showstopper to long-duration spaceflight and habitation. According to the NRC study, current "models predict that crewed missions beyond LEO would be limited to three months or less because of adverse health impacts, either during the mission or during a crewmember's lifetime." That limit is less than half the time needed to reach Mars. Advanced protection methods must be assessed before embarking on crewed interplanetary missions, and that means we still have plenty of homework to do. As the NRC found:

> Without the collection of in-situ biological data to support the development of appropriate models, as well as the development of new sensors, advanced dosimetry instruments and techniques, solar event prediction models, and radiation mitigating designs, extended human missions to the Moon, Mars, or near-Earth asteroids (NEAs) may be beyond acceptable risk limits for both human health and mission success.

A relatively inexpensive attempt to address this problem was launched with the Mars Science Laboratory in November 2011. An instrument weighing less than four pounds called the Radiation Assessment Detector was attached to the *Curiosity* rover to characterize the radiation environment during the cruise phase and on the Martian surface. More investigations like this are needed, perhaps employing a series of small satellites sent to a deep space location with each one testing a different shielding concept. Lunar orbit would be a good choice, since the Moon will be an early target for human habitation, and lunar-observing instruments could be added to increase the utility of the mission.

Another ISS limitation is that it tells us nothing about how to function on a planetary surface with gravity that is a fraction of Earth's. The weightless environment of the ISS needs to be supplemented by a variable gravity facility to determine if spinning spacecraft make sense for long flights, and if so, at what gravity level. A one-g environment may not be necessary to maintain good health and full functionality, and there are technical advantages to designing a spacecraft for lower spin rates. The facility can simulate planetary gravity environments that we expect to encounter, such as one-sixth-g on the Moon or one-third-g on Mars.

The NRC technology priorities constitute a sensible list from a perspective focused on human and robotic missions to conduct on-site and remote science investigations throughout the solar system. But some of the big questions that desperately need answers may receive cursory attention if we choose a spaceflight approach that only looks beyond the Moon and is overly eager to put human footprints on alien worlds. From the perspective of executing Cislunar-Next, the list would be different, although there would be some overlap.

Under Cislunar-Next, developing the technologies and experience that enable efficient, sustainable, expandable space operations would be top priority. Let's consider some of the priority items for Cislunar-Next that might receive inadequate attention under other approaches that simply aim at stepping-stones to Mars and other solar system destinations.

On-orbit servicing. Repair and refueling of satellites was touched on earlier in the discussion of proximity operations. If we're serious about living and working in space for the long haul, we're not going to discard our hardware every time it breaks down or runs out of juice. We're going to learn how to refill its tank, replace its gaskets, give it a tune-up, extend its life, and upgrade its capabilities. This has to become routine, unlike the elaborate and expensive Hubble Space Telescope repair missions. As much as possible, we'll do the job with automated or tele-operated robots. Demonstrating the robotics should be straightforward—satellite servicing will be done in a structured environment (human-made devices working on each other), and tele-operation is an option throughout cislunar space because everything is less than a light-second away.

In March 2012, NASA began a series of robotic servicing demonstrations aboard the ISS intended to last two years. The agency created a set of custom tools for the station's Canadian robot arm, and a spacecraft mock-up with valves, wires, seals, and thermal blankets like those on commercial satellites. The project's goal is to practice techniques for on-orbit refueling of satellites that were never designed to be refueled. Meanwhile, the Defense Advanced Research Projects Agency began a program called Phoenix, in which a robotic servicer would attempt to recycle parts (such as antennas) from defunct communications satellites. Programs like these are precisely what's needed, but their success will be measured by the extent of follow-on work and the ability to attract private-sector operators to join this activity.

Standardization. If retrieval, repair, and refueling of space hardware is to occur, it will be enabled and assisted by standardization. The difficult part will be getting manufacturers to redesign their space hardware to be serviced by robots using common interfaces. Difficult, but achievable, since only a small number of organizations worldwide routinely produce large satellites, and they're already doing standardization to integrate their payloads with multiple launch vehicles from several countries. By comparison, it would seem a simple matter to settle on standard grappling fixtures so that satellites can be captured safely and efficiently by service vehicles. Also needed are standard ports for fuel and other fluids, electric power, and data transfer. Replacement of old or malfunctioning parts could be done with modular components—Orbital Replacement Units (ORUs), as NASA calls them. Once these standards are in place, they can be carried over to modular assembly of large platforms. Space development guru John Mankins of Artemis Innovations has said that we need to find the space equivalent of USB. (Will that acronym someday stand for Universal Spacecraft Bus?) If we can do that, we'll be taking a key step toward a new space economy.

Inter-orbital transportation. When NASA first envisioned the Space Transportation System in the 1970s, it was going to be more than just the shuttle. Another element was called the Orbital Transfer Vehicle (OTV), a reusable upper stage (or "space tug" as some called it) that would have taken payloads from the shuttle's orbit to higher orbits. Like the cancelled OMV mentioned earlier, this was another missed opportunity to move toward Stage Two. In the coming decades, in-space transportation needs to have a renaissance comparable to the experience of automobiles, ships, and aircraft in the 20th century. This will produce a wide variety of craft that are sized and specialized for particular tasks. Just as cars, trucks, ships, and aircraft come in an assortment of shapes and sizes, so will future space vehicles that travel between LEO, GEO, lunar orbit, and Lagrange points. They will drop off and retrieve many kinds of payloads, and will carry robots and humans to locations where they're needed.

Fuel storage. An earlier chapter briefly mentioned the possibility of using on-orbit fuel depots. An internal NASA study in 2011 assessed the use of a fuel depot in low Earth orbit that would fill up the final stages of missions bound for the Moon or points beyond. The study's rough cost estimates purported to show significant cost savings compared to using a government-developed heavy-lift rocket that carried all of its fuel at launch. This may be true, but further analysis is required to determine whether the depot concept makes economic sense in the broader scheme of space development. For example, if deep space missions using the depot occur on a regular basis (multiple times per year), this would seem to be a sensible investment. But if the orbital gas station's only customers are NASA's own missions, flying an average of less than once a year (as the agency's study assumed), it's difficult to see how NASA would justify such an underutilized facility. The demands of building, deploying, and operating the depot would be considerable. A sizable platform would have to be designed with plenty of shielding for the LEO debris environment, thermal control to handle the 90-minute day/night cycles in LEO (which would be particularly challenging if the propellants were cryogenic), and its own propulsion system to occasionally boost its orbital altitude and maneuver to avoid collisions. Techniques and systems for on-orbit propellant transfer and measurement still need to be refined. Add to that the cost of routine commercial launches to top off the tank.

The orbiting fuel depots developed for Cislunar-Next would not necessarily be in LEO, nor would they always get their fuel supplies from Earth. They would be located where the action is, be it in LEO, GEO, lunar orbit, or Lagrange points. Some of their supplies could come from the hydrogen and oxygen extracted from lunar or asteroidal ice. Their best customers would come from the inter-orbital traffic throughout cislunar space (for example, satellite servicing bots and reusable orbital transfer stages) with less-frequent visits from deep space missions needing a fill-up on their way out. The customer list would encompass the world's spacefarers, public and private—not just a single agency from a single country. This is another place where standardization would be essential. Just like terrestrial gas stations, all manner of space motorists should be able to pull up to the pump, swipe their credit card, and neatly fit the dispenser nozzle into their own tank.

Materials processing in space (MPS). An important component of the space economy will be microgravity materials processing. The reaction of some people with long memories may be, "Been there, done that, didn't work." But that quick dismissal would be a mistake. It's true that since the 1980s, there's been plenty of attention devoted to MPS in attempts to discover unique properties and take advantage of processes and conditions not available on Earth. Plenty of attention, but not enough on-orbit research. Access to lab space on orbit has been extremely hard to come by, and very expensive to use, so the basic research phase of this activity has stretched out with no apparent end in sight. So we still haven't answered questions about which processes result in useful products, how those processes might be scaled up to industrial production levels, and whether any of this can be turned into a business plan with a happy ending. It is vital that we answer these questions. There won't be much of an economic future in space if all materials processing and manufacturing has to be done back on Earth.

Extraterrestrial resources. MPS is related to another set of important questions on in-situ resource extraction and utilization. Science fiction writers and real-world space planners have been talking about this for decades, but we still don't have definitive answers. Can we mine the Moon and asteroids for minerals? How would terrestrial mining methods need to be modified for the task? Should materials be refined on site, or in a separate orbiting facility? What kinds of final products will use these materials? Will the products only be used in space, or will they be marketable on Earth?

Much has been made of the strong evidence that large deposits of water ice exist in permanently shadowed craters near the poles of the Moon. That's great news, but those deposits still need to be located precisely and their extent has to be estimated more accurately. Then we need to figure out how to "mine" the ice. The extremely low temperature will make it as hard as the rock around it, so it's going to take specially designed machines—not a couple of astronauts with ice picks—to get it out. Once extracted, it must be transported to a facility for processing, to turn it into potable water or to separate the hydrogen and oxygen components to supply fuel, oxidizer, or breathable oxygen. All of this must be demonstrated before we can count on lunar ice as a critical element in the cislunar infrastructure.

Energy collection and distribution. Adequate power needs to be available at widely dispersed locations. What should be the balance between solar, nuclear, and fuel cell power sources? What particular system design is best in each of these categories? We won't know the answers to these questions until we get fully engaged in Cislunar-Next.

Long before we feel the need to import minerals to Earth from space, we may find ourselves looking skyward for a source of abundant clean energy. Solar power platforms can collect the Sun's energy and beam it to where it's needed, whether that's other orbiting platforms, lunar outposts, or the surface of the Earth. But we have yet to conduct a pilot project to demonstrate this in space, which must be followed up by efforts to scale it up to industrial size. If we're developing techniques for lunar materials extraction and processing at the same time, the solar platforms could be the driver for exploring how much we can build in space with materials that don't have to be lifted up from Earth.

A solar power satellite (SPS) concept called the Arbitrarily Large Phased Array (ALPHA). In general, SPS platforms would collect energy in GEO and beam it to one or more receiving antennas on Earth using microwaves or lasers. ALPHA would be constructed from thousands of small mass-produced components.
Source: SPS-ALPHA concept by John Mankins; graphic by Mark Elwood, SpaceWorks Enterprises, Inc.

Other in-space utilities. If operations throughout cislunar space become routine, there will be a need for dedicated communications and navigation services like the ones we're used to on Earth and in LEO. Existing services are all aimed at serving Earth, so additional systems are needed to serve other parts of cislunar space. That doesn't mean the Moon needs a 30-satellite constellation for navigation, or that lunar orbit should be filled with communications satellites comparable to those in GEO. But it does recognize that operations on the scale being discussed here can no longer depend on research facilities like NASA's Deep Space Network to provide all that is needed. Whatever architecture is chosen to provide "comm and nav" to cislunar space, it should be designed to operate as transparently as possible to its users without routing everything through Earth and a room full of mission controllers.

Another essential utility is space weather. Human crews living and working in high orbits or on the Moon need timely warnings and analyses of solar activities that could have dire effects on their health and their technical systems. Ideally, they should have real-time links to the warning systems to avoid any delays in alerts from Earthbound observers. Future human activities spread across cislunar space may not have the luxury of around-the-clock monitoring by teams of technicians as the ISS does today.

This discussion of priorities is not intended to be an exhaustive list of the things we can or should do in cislunar space, but it does show that even after a half-century of spaceflight, there are plenty of things we don't know and can't do in our own celestial backyard. The point of this is not to pave the way for countless thousands of people to work in space as quickly as possible; in fact, the more we can do with automated systems, the better. The point is to identify the hurdles that need to be overcome to vastly increase humanity's resources, capabilities, and knowledge.

Measuring progress

For many years, aerospace engineers have been measuring the sophistication of technical developments using Technology Readiness Levels (TRLs). NASA's definitions of the nine levels are shown in the table below. They trace the evolution of a system concept from the establishment of its basic technical principles (TRL 1) through its successful use in mission operations (TRL 9). This should not be taken to imply steady progress at some predictable pace. The stages from TRL 5 through TRL 7 have been called "the valley of death" because many technologies go in but never come out. That's the point at which everything gets harder. The research gets more expensive—often a lot more. The hardware needs to be scaled to the proper size and demonstrated in a real-world environment, not just in the lab. Flight opportunities are essential, but may be difficult or impossible to obtain. The project needs champions who can skillfully guide it through the competition for scarce resources. Many of the critical systems needed to conduct efficient, sustainable operations in cislunar space have not yet reached the valley of death, or are currently mired in it. Some have been stuck there for a long time, awaiting funding, or a champion, or just a chance to prove themselves in spaceflight. The concerted effort of Cislunar-Next is needed to get them unstuck.

Technology Readiness Levels

TRL 1	Basic principles observed and reported
TRL 2	Technology concept and/or application formulated
TRL 3	Analytical and experimental critical function and/or characteristic proof of concept
TRL 4	Component and/or breadboard validation in laboratory environment
TRL 5	Component and/or breadboard validation in relevant environment
TRL 6	System/subsystem model or prototype demonstration in a relevant environment (ground or space)
TRL 7	System prototype demonstration in a space environment
TRL 8	Actual system completed and "flight qualified" through test and demonstration (ground or space)
TRL 9	Actual system "flight proven" through successful mission operations

Chapter 12

Not the end

Human history becomes more and more a race between education and catastrophe.—H.G. Wells, British science fiction writer and social commentator

It may seem surprising in an age when more people have access to more information than at any time in human history, but most Americans get the majority of their information on public policy issues from just two sources: their elected officials and the general media. When the subject is as complicated as space or other technology issues, it's possible and often likely that those sources will misunderstand or misrepresent the issue. Members of the public typically will not seek out scientific and technical expertise (for example, by attending a lecture, taking a class, or reading specialized books, periodicals, or websites) to resolve discrepancies and fact-check their usual sources. Instead, they will labor under misconceptions or incomplete awareness of the essential facts and circumstances. In a democracy, this can lead to bad decisions, or at least decisions that are delayed too long. This is the modern-day version of the race between education and catastrophe that aroused the concern of H.G. Wells nearly a hundred years ago.

Many pundits exemplify this education problem when they speculate that the space age is over. Granted, the events of the past few years, including some I've chronicled here, have provided evidence that could support that point of view. But anyone willing to take a deep dive into the history and future possibilities of space exploration and development will recognize its richness and potential. Despite being a long-time observer of space-related activities that are often sluggish and sometimes misguided, I remain a loyal follower of the school of You Ain't Seen Nothin' Yet.

Getting on track

Sometimes we take actions that seem to be a deliberate step backward. In *Choice, Not Fate*, I lamented the termination of the NASA Institute for Advanced Concepts (NIAC), which was shut down in August 2007 for reasons of budgetary expediency. For comparatively little money and with a very small staff, this organization had supported the kind of work that will enable our long-term future in space. Established in February 1998, it offered small grants to universities and businesses studying innovative technologies that could be realized in 10 to 40 years. During its short existence, NIAC awarded 126 study grants, plus 42 follow-on contracts, for a total investment of $27.3 million (pocket change by government standards) to discover new scientific and technical talent and encourage concept development for tomorrow's challenges. Here is a sampling of the topics studied during that 10-year period that have applicability to Cislunar-Next:

- Tether-based transportation throughout cislunar space
- Space elevators for ascent to orbit from the Earth and the Moon
- Use of electromagnetic fields for radiation shielding, especially on the Moon
- Lunar colony architecture, including large self-deployed habitats and structures
- Self-replicating lunar factories
- Solar power satellite cell fabrication in space
- Lunar ice recovery

The products of the entire series of NIAC studies have been preserved in an online database accessible to all. (See the bibliography for the Internet link.)

I'm happy to report that NIAC is back. In 2011, it returned with the same acronym but a slightly different name: the NASA Innovative Advanced Concepts Program. Its mandate is the same as before. In terms of Technology Readiness Levels, it will support research at TRL 1, 2, and possibly 3. The new effort made 30 awards in its first phase, and expected to make several more in its second phase in 2012. First-phase awards included studies on topics applicable to Cislunar-Next such as:

- Space debris elimination
- Innovative power generation concepts
- Solar power satellite design
- Planetary surface mobility
- Radiation shielding materials and methods

It should be obvious from the topics I've highlighted here, and even more so from a thorough perusal of the complete set of topics posted online, that NIAC is making an important contribution to the priorities identified in the previous chapter—but only at low levels of technology readiness. The hard part is advancing all the way up to operational systems in an environment that is resource constrained and includes the countervailing forces of short-term thinking, partisanship, and parochialism.

These forces have been around throughout the space age and have always been influential. However, in recent decades, they have increased the danger that the vital enterprise of space exploration and development will become little more than a hobby shop for builders of rockets and spacecraft, and a jobs program for a handful of congressional districts. If that's the case, then the United States has lost something of great value: one of its most potent tools for building a better future.

The U.S. probably will always retain the ability to build space hardware, but that's not the same as building a future. NASA's efforts to develop the Space Launch System and the Orion capsule, the origins of which were discussed on these pages, provide a prominent example of the difference. On the positive side, if SLS and Orion complete their development and testing and become operational systems, they could constitute a safe, reliable launch system that sets a new standard for heavy-lift access to space. This could be very useful in achieving milestones in Cislunar-Next. However, there are still a couple of concerns.

The first concern is whether a big new launch system based on legacy technologies adequately addresses the traditional factors of cost, schedule, and performance. Recent generations of launch vehicles, in the U.S. and around the world, have demonstrated considerable success in advancing aspects of performance such as capacity, precision of orbital insertion, and overall reliability. Cost and schedule, on the other hand, have proven far more resistant to efforts at improvement. At this point, there's no reason to believe that the experience with SLS will be any different. The cost to orbit per unit of payload could be as high or higher than what we have today aboard our largest boosters, and the flight schedule may be no more predictable or flexible than with today's rockets. If that turns out to be the case, then what have we gained for our great effort and investment?

The second concern gets more to the heart of building a better future: if we don't advance to a new level in our strategic planning and long-term commitment, SLS/Orion will be all dressed up with nowhere to go. It would be overkill to use a big, expensive rocket like SLS to service the space station, as that duty can be handled by other means, including commercial services. Also, SLS is not scheduled to be fully operational until 2021, and there's no guarantee that the space station will still be functioning at that time (although I hope it will be).

Most participants in the space community recognize that SLS/Orion lacks a well-defined purpose beyond providing jobs in certain regions of the country. This makes for some uncomfortable conversations with people who really want to understand the rationale behind the launcher program, and more generally, the future of U.S. human spaceflight.

"What's the purpose of SLS/Orion?"
"To take humans beyond LEO."
"To do what?"
"Go to the Moon, asteroids, and Mars."
"For what purpose?"
"To do science, and inspire our youth."
"So let me get this straight. SLS/Orion is the first installment on a bazillion-dollar program to send people on high-risk missions to do science and inspire youth, both of which are already being accomplished far more affordably by robots, at no risk to humans."
"Uh . . . well . . . it's our destiny."

I've seen face-to-face and online variations of this conversation, and I expect that they take place all the time, repeatedly demonstrating the failure of the destination-driven rationale for human spaceflight.

If the U.S. is intent on building SLS/Orion, then it must be given a purpose. The NASA Advisory Council, an independent group that advises the NASA administrator, took up this question at its March 2012 meeting, but disappointingly, came up with precisely the wrong recommendation:

> . . . now is the time to pick a specific destination in order to focus the NASA, international agencies and contractor

teams on a specific destination, such as Mars . . . We believe that a focused mission with a specific end objective, as has been the case for over 50 years for Human Spaceflight Programs, would also greatly benefit the NASA workforce, current and future domestic and international partners and the public stakeholders.

This advice "to pick a specific destination in order to focus . . . on a specific destination" misrepresents the fact that "a focused mission with a specific end objective" was an approach used only once in the U.S. human spaceflight program, not throughout the past 50 years. The Council seems to believe that our thinking a half-century ago was good enough then, so it's good enough now. They assess the consequence of not taking their recommended action as follows: "Without selecting a mission we will delay a human flight to a destination." This is a classic example of mistaking destinations for goals, and the problem is compounded by the emphasis on near-term job creation rather than on a lasting, value-generating rationale. Our metrics for success should not be how quickly we get to Mars or how many people we have living in space; rather, we should be measuring how much we're gaining in capabilities and knowledge, leading to increased prosperity, global solutions, and discovery.

We can do a better job of giving a purpose (or better yet, multiple purposes) to SLS/Orion and other evolving space systems. We can carry out missions in support of Cislunar-Next. Build things, like microgravity laboratories, manufacturing facilities, lunar outposts, and solar power satellites. Create or enhance capabilities, like satellite servicing and the harvesting of extraterrestrial energy and material resources. In other words, do things that generate scientific, economic, and societal value, bringing quantitative and qualitative benefits to Earth and justifying the continued exploration and development of space.

If SLS/Orion is not the right vehicle at the right price to support these missions or others deemed worthy, then we should reconsider it before expending any more resources on it. The opportunity cost—inability to sufficiently investment in other critical space ventures—may be too high. In an era of tight budgets, SLS/Orion could become the tail wagging the dog at NASA, soaking up an inordinate amount of funding and management attention throughout its development and operations, at a time when emerging private-sector capabilities may make this unnecessary.

By the time we get to 2021, when SLS/Orion is scheduled to be fully operational, many (perhaps all) of the program's Capitol Hill champions won't be there any longer, and we will have a different president. Will the decision-makers of that time be adding up the costs and asking why we're doing this? Are we in for a lot more years of shifting priorities and cancelled projects? It doesn't have to be that way if we can come together on a strategy with purpose like Cislunar-Next and fully engage partners in the international and commercial sectors.

In order to move to the next level of maturity in our quest to become Stage Two spacefarers in the coming decades, we need to embrace what's been created so far in the space age—much of it through NASA's influence—by taking advantage of growing cross-sector abilities. Offload operational and routine functions to the private sector and other entities that are increasingly equipped to handle them, and let NASA focus on the scientific research and technology development activities at which it excels, and which otherwise would receive inadequate attention.

Major themes

For much of the time since the end of the Apollo era, participants in the space community and members of the interested public have been frustrated by the inability to choose a coherent path for space exploration and development, make a commitment to its implementation over the long term, and fund it appropriately. Considering just the critical area of access to space, the community has lost count of how many launcher development programs have been started and terminated without producing a working vehicle. (There's a joke that says 70 percent of the Earth is covered by water, and the other 30 percent is covered by launch studies.) You can't blame these people for expressing their aggravation by exclaiming, "Just pick something and stick with it!" Despite this, I've been suggesting throughout this book that more change is in order.

Why would I do such a thing? For the short answer, I'll repeat what I said at the beginning: in this high-risk, high-cost, long-term venture, it's more important to get it right than to do it fast. For a longer answer, I'll summarize the major themes presented here:

- **Three stages of space development**. We're currently in Stage One, still in training as we have been for the past half century. In Stage Two, we will create an industrial park that extends throughout cislunar space. Once we've accomplished that milestone will we be ready to set new goals leading to Stage Three, settlement of cislunar space and human movement beyond it to the rest of the solar system.

- **Spacefaring goals for the current era.** It's difficult to aim for something when the ground seems to be constantly shifting beneath your feet. Nonetheless, if we intend to reach Stage Two, our spacefaring goals should be characterized by the establishment and sustainable operation of space infrastructure that creates economic value, societal benefits, and new knowledge. This is something that partisans of all stripes should be able to embrace, because it has much to offer to a broad swath of the ideological spectrum. Rather than building launch systems in the hope that their purpose will become clear later, we should immediately formulate and enact a comprehensive strategy for the development of cislunar space. This will build the fundamentals needed to venture beyond cislunar space with human crews when the time is right and the capabilities are in hand. Human spaceflight is a component of spacefaring, but success should not be measured by the number of people living in space. Ongoing accomplishment in science, security, and economic development—all of which will be highly dependent on robotic systems—should be key determinants of the pace of human migration.
- **Sequential vs. parallel development of space infrastructure.** We can't reach critical mass if infrastructure elements are introduced one at a time and don't survive long enough to assemble into an integrated whole. Broader participation by the private sector is required to achieve parallel development of infrastructure elements and long-term commitment to operations.
- **Public-private collaboration.** The public sector has an important role to play in performing basic research, buying down risk, and otherwise creating an environment that will encourage private-sector entry. History demonstrates that big undertakings in modern American society are collaborations between the public and private sectors. It is counterproductive—and unsupported by historical evidence—to argue that either sector should go it alone. History disproves notions that only governments can or will develop space, that only the private sector is capable of creative solutions to problems, or that regulations always harm industry.

- **The importance of going mainstream.** Unless space can become more to society than just a prestige activity and a better path for relaying electromagnetic signals, it will be difficult for the U.S. civil space program to rise above being a pawn in a political environment characterized by partisanship, parochialism, and short-term thinking. Controversy over public and private-sector roles will be an inhibitor until those roles become more firmly established. When that happens, long-term strategy for space can get beyond the problem of being overshadowed by short-term tactical interests like job creation.

Instead of getting all dreamy about what we *could* do, or becoming irritated as pessimists carry on about what we *can't* do, or anguishing over what other nations *might be trying* to do, let's just start doing our homework. Can we repair, refuel, and recycle all those expensive satellites we put into orbit rather than writing them off as space debris? Can we usefully mine the Moon or asteroids? Can we do large-scale collection of solar energy and distribute it to users in space and on Earth? If so, what will it take to make all of this a going concern? We can sit at the bottom of Earth's gravity well, enjoying the intellectual stimulation that such contemplation provides, or we can go out and try it for real.

Acronyms

ASAP	Aerospace Safety Advisory Panel
ASAT	anti-satellite weapon
CAA	Civil Aeronautics Authority
CAB	Civil Aeronautics Board
CAIB	Columbia Accident Investigation Board
DoC	Department of Commerce
ESA	European Space Agency
FAA	Federal Aviation Administration
FY	fiscal year
GAO	Government Accountability Office (formerly General Accounting Office)
GEO	geosynchronous equatorial orbit
ICC	Interstate Commerce Commission
ISS	International Space Station
JPL	Jet Propulsion Laboratory
LEO	low Earth orbit
MAA	Manufacturers Aircraft Association
MPCV	Multi-Purpose Crew Vehicle
MPS	materials processing in space
NACA	National Advisory Committee for Aeronautics
NASA	National Aeronautics and Space Administration
NGO	non-governmental organization
NIAC	NASA Innovative Advanced Concepts Program (formerly NASA Institute for Advanced Concepts)
NOAA	National Oceanic and Atmospheric Administration
NRC	National Research Council
NSC	National Security Council

OECD	Organization for Economic Cooperation and Development
OMB	White House Office of Management and Budget
OMV	Orbital Maneuvering Vehicle
ORU	orbital replacement unit
OSTP	Office of Science and Technology Policy (the president's science advisor)
OTV	Orbital Transfer Vehicle
R&D	research and development
SLS	Space Launch System
TRL	Technology Readiness Level
U.N.	United Nations

Bibliography

Chapter 1: Are we there yet?

Cleveland *Plain Dealer* editorial. "U.S. cannot responsibly avoid a significant investment in its space program," September 24, 2009 (http://www.cleveland.com/opinion/index.ssf/2009/09/us_cannot_responsibly_avoid_a.html).

Organization for Economic Cooperation and Development. "Infrastructure for 2030: Mapping Policy for Electricity, Water, and Transport," 2007 (http://www.oecd.org/dataoecd/61/27/40953164.pdf).

Review of Human Spaceflight Plans Committee (Augustine Committee). "Seeking a Human Spaceflight Program Worthy of a Great Nation," October 22, 2009 (http://www.nasa.gov/pdf/396093main_HSF_Cmte_FinalReport.pdf).

Space Settlement Act of 1988. Public Law 100-685, Section 217 (part of the NASA Authorization Act for Fiscal Year 1989).

Space Task Group, "The Post-Apollo Program: Directions for the Future," September 1969 (http://www.hq.nasa.gov/office/pao/History/taskgrp.html).

Chapter 2: Will we ever get there?

Aftergood, Steven. "Poisoned plumes: Across the U.S., environmentalists are protesting against rocket launches. Toxic exhaust fumes from rockets packed with solid propellant attract the greatest concern." *New Scientist*, September 7, 1991 (http://space.newscientist.com/channel/space-tech/space-shuttle/mg13117854.400).

Bergaust, Erik. *The Next Fifty Years in Space* (New York: Macmillan, 1964).

The Economist editorial, "The end of the Space Age: Inner space is useful. Outer space is history," June 30, 2011 (http://www.economist.com/node/18897425?story_id=18897425&fsrc=rss).

Diamond, Jared. *Collapse: How Societies Choose to Fail or Succeed* (New York: Penguin Group, 2011).

Environmental Working Group. "Lab Tests Show Traces of Rocket Fuel in Lettuce," April 28, 2003 (http://www.ewg.org/news/lab-tests-show-traces-rocket-fuel-lettuce).

Gagnon, Bruce. "NASA Plans Moon Base to Control Pathway to Space," December 13, 2006 (http://www.space4peace.org/articles/nasa_moon_base.htm).

Global Network Against Weapons & Nuclear Power in Space. "NASA & Pentagon Surveillance of Global Network," April 20, 2005 (http://www.space4peace.org/reports/nasa_pentagon_spy_on_gn.htm).

Greason, Jeff. Keynote speech at the International Space Development Conference, May 21, 2011 (http://moonandback.com/2011/06/12/jeff-greason-a-settlement-strategy-for-nasa/).

Grossman, Karl. "Bush Opens Outer Space to Combat," October 25, 2006 (http://www.space4peace.org/articles/bush_opens_space_to_combat.htm).

Lambright, W. Henry. "Managing America to the Moon: A Coalition Analysis" in *From Engineering Science to Big Science: The NACA and NASA Collier Trophy Research Project Winners*, ed. Pamela E. Mack (Washington, DC: NASA SP-4219, 1998), p. 209 (http://history.nasa.gov/SP-4219/Chapter8.html).

Mann, Charles C. *1491: New Revelations of the Americas Before Columbus* (New York: Vintage Books, 2005; second edition 2011).

Ross, Martin, Michael Mills, & Darin Toohey. "Potential climate impact of black carbon emitted by rockets," Geophysical Research Letters, Vol. 37, December 2010 (http://www.agu.org/pubs/crossref/2010/2010GL044548.shtml).

U.S. Office of Technology Assessment. *Solar Power Satellites*, August 1981, pp. 179-224 & 275-288 (http://www.princeton.edu/~ota/disk3/1981/8124_n.html).

Zubrin, Robert. "NASA Needs a Destination," *Space News*, February 22, 2010, p. 19 (http://www.spacenews.com/commentaries/100222-nasa-needs-destination.html).

Chapter 3: Hope, change, and the space program

Aviation Week & Space Technology editorial. "Spaceflight and Mr. Augustine: The Prudence Reflected by a Human-Spaceflight Study May Serve Obama and the U.S. Well," May 18, 2009, p. 62 (http://www.aviationweek.com/publication/awst/loggedin/AvnowStoryDisplay.do?fromChannel=awst&pubKey=awst&channel=awst&issueDate=2009-05-18&story=xml/awst_xml/2009/05/18/AW_05_18_2009_p62-140946.xml&headline=The+Prudence+Reflected+By+A+Human-Spaceflight+Study+May+Serve+Obama+And+The+U.S.+Well).

Bush, George W. "U.S. Space Exploration Policy," National Security Presidential Directive (NSPD)-31, January 14, 2004.

Chang, Kenneth. "Review Panel Hears Rival Plans for New Spaceflights," *New York Times*, June 18, 2009 (http://www.nytimes.com/2009/06/18/science/space/18nasa.html).

Griffith, Rep. Parker (AL). "Statement of Representative Griffith on Full Release of Augustine Report," press release, October 22, 2009 (http://griffith.house.gov/index.cfm?sectionid=24&sectiontree=23,24&itemid=390#).

Lizza, Ryan. "The Obama Memos," *The New Yorker*, January 30, 2012 (http://www.newyorker.com/reporting/2012/01/30/120130fa_fact_lizza?printable=true&currentPage=all).

NASA. "Charter of the Review of U.S. Human Space Flight Plans Committee," June 1, 2009.

Obama, Barack. "Advancing the Frontiers of Space Exploration," campaign fact sheet, August 16, 2008 (http://www.spaceref.com/news/viewsr.html?pid=28880).

Posey, Rep. Bill (FL). "NASA Needs More Funding to Close the Gap and Maintain America's Leadership in Space Exploration," press release, October 22, 2009 (http://posey.house.gov/News/DocumentSingle.aspx?DocumentID=151015).

Review of Human Spaceflight Plans Committee (Augustine Committee). "Seeking a Human Spaceflight Program Worthy of a Great Nation," October 22, 2009 (http://www.nasa.gov/pdf/396093main_HSF_Cmte_FinalReport.pdf).

Shelby, Sen. Richard (AL). "NASA and the Future of Human Spaceflight," *Congressional Record*, October 21, 2009, p. S10614 (http://www.spavcepolicyonline.com/pages/images/stories/Shelby_Senate_Floor_Oct_21_09.pdf).

Smith, Marcia. "Congressional Reaction to Augustine Committee Report Suggests Long Road Ahead, *Space Policy Online*, October 22, 2009 (http://www.spacepolicyonline.com/pages/index.php?option=com_content&view=article&id=499:congressional-reaction-to-augustine-committee-report-suggests-long-road-ahead&catid=67:news&Itemid=27).

U.S. Congress. "Consolidated Appropriations Act, 2008," Public Law 110-161, Division B, Title III, December 26, 2007.

White House Press Release. "President Bush Announces New Vision for Space Exploration Program," speech at NASA Headquarters, Washington, DC, January 14, 2004.

Chapter 4: 2010: The year we made conflict

Armstrong, Neil, James Lovell, and Eugene Cernan. Statement on President Obama's proposed changes to NASA's human spaceflight program, April 13, 2010 (http://www.msnbc.msn.com/id/36470363/ns/nightly_news/).

Astronauts' letter to Sen. Barbara Mikulski (Chair, Senate Appropriations Subcommittee on Commerce, Justice, Science and Related Agencies) supporting commercial crew transport. Signed by 22 shuttle and two Apollo astronauts, July 14, 2010 (http://www.spaceref.com/news/viewpr.html?pid=31239)

Aviation Week & Space Technology editorial. "A Cloudy Vision," February 8, 2010, p. 50.

Barbee, Brent W. (ed.). "Target NEO: Open Global Community NEO Workshop Report," George Washington University, Washington, DC, July 28, 2011.

Boyle, Alan. "Private spaceflight goes public," *MSNBC Cosmic Log*, February 01, 2010 (http://cosmiclog.msnbc.msn.com/archive/2010/02/01/2191461.aspx).

Carreau, Mark & Frank Morring, Jr. "Space Rocks: Obama's call for human asteroid mission spurs work on concepts," *Aviation Week & Space Technology*, Sep 6, 2010, p. 36 (http://www.aviationweek.com/publication/awst/loggedin/AvnowStoryDisplay.do?fromChannel=awst&pubKey=awst&channel=awst&issueDate=2010-09-06&story=xml/awst_xml/2010/09/06/AW_09_06_2010_p36-251431.xml&headline=Concepts+For+Robotic%2C+Human+Asteroid+Missions+Are+Proposed).

Columbia Accident Investigation Board (CAIB). Letter from five of the 13 members to Sen. Barbara Mikulski (Chair, Senate Appropriations Subcommittee on Commerce, Justice, Science and Related Agencies) supporting commercial crew transport, July 12, 2010 (http://www.spaceref.com/news/viewsr.html?pid=34471).

Hall, Ralph. "Ranking Member Hall Questions President's Decision to Cancel Constellation," press release, February 1, 2010 (http://gop.science.house.gov/Pressroom/Item.aspx?ID=219#).

Hatch, Orrin, Bob Bennett, Rob Bishop, & Jason Chaffetz. "Killing Ares Will Kill the Dream," *Aviation Week & Space Technology*, Apr 19, 2010, p. 66 (http://www.aviationweek.com:80/publication/awst/loggedin/AvnowStoryDisplay.do?fromChannel=awst&pubKey=awst&channel=awst&issueDate=2010-04-19&story=xml/awst_xml/2010/04/19/AW_04_19_2010_p66-219941.xml&headline=U.S.+Needs+Ares+To+Maintain+The+Skills+For+Space+And+Strategic+Deterrent+Programs).

Hsu, Jeremy. "Few Candidate Asteroids for Proposed Manned Mission by 2025," *Space News*, September 6, 2010, p. 12 (http://www.spacenews.com/civil/100906-few-candidate-asteroids-manned-mission.html).

Hutchison, Sen. Kay Bailey. "After 50 years of NASA, We Must Not Leave Space," *Houston Chronicle*, March 6, 2010 (http://hutchison.senate.gov/opedNASA_HC_030610.html).

Klamper, Amy & Debra Werner. "NASA's New Direction Drawing Fire from House and Senate Lawmakers," *Space News*, March 1, 2010, p. 1 (http://www.spacenews.com/policy/100226-nasa-new-direction-drawing-fire-lawmakers.html).

Klamper, Amy. "House Appropriators Grill Obama's Science Adviser on NASA Plan," *Space News*, March 1, 2010, p. 6 (http://www.spacenews.com/policy/100225-house-appropriators-grill-obama-science-adviser.html).

Logsdon, John M. "The Obama Plan: Risks Worth Taking," *Space News*, March 15, 2010, p. 27 (http://www.spacenews.com/commentaries/100315-obama-plan-risks-worth-taking.html).

Logsdon, John M. "The End of the Apollo Era—Finally?" *Space News*, July 5, 2010, p. 19 (http://www.spacenews.com/commentaries/100630-blog-end-apollo-era-finally.html).

Morring, Frank. "Commercial Route," *Aviation Week & Space Technology*, February 8, 2010, p. 20.

Moskowitz, Clara. "Misconceptions Surround White House NASA Plan, Experts Say," *Space News*, June 28, 2010, p. 10 (http://www.spacenews.com/civil/100628-misconceptions-white-house-nasa-plan.html).

National Research Council. "An Interim Report on NASA's Draft Space Technology Roadmaps," National Academies Press, 2011 (http://www.nap.edu/catalog.php?record_id=13228).

Norris, Guy. "Close Encounters: Group crafts risk-warning and management blueprint to counter asteroid threat," *Aviation Week & Space Technology*, November 28, 2011, p. 51 (http://www.aviationweek.com:80/publication/awst/loggedin/AvnowStoryDisplay.do?fromChannel=awst&pubKey=awst&channel=awst&issueDate=2011-11-28&story=xml/awst_xml/2011/11/28/AW_11_28_2011_p51-395551.xml&headline=Asteroid+Action+Plans+Form+For+Planetary+Defense).

Obama, Barack. "Remarks by the President on Space Exploration in the 21st Century," Kennedy Space Center, Florida, April 15, 2010 (http://www.whitehouse.gov/the-press-office/remarks-president-space-exploration-21st-century).

Obama, Barack. "National Space Policy of the United States of America," June 28, 2010 (http://www.whitehouse.gov/sites/default/files/national_space_policy_6-28-10.pdf).

Orlando Sentinel. "Griffin, NASA luminaries urge Obama to change space policy," April 12, 2010 (http://blogs.orlandosentinel.com/news_space_thewritestuff/2010/04/griffin-nasa-luminaries-urge-obama-to-change-space-policy.html).

Shelby, Senator Richard C. "Obama plan would destroy U.S. space supremacy," statement to the Senate Appropriations Subcommittee on Commerce, Justice, Science, and Related Agencies, April 22, 2010.

Space News editorial. "Change Springs Eternal," February 8, 2010, p. 16 (http://www.spacenews.com/commentaries/10028-change-springs-eternal.html).

Space News editorial. "Latest NASA Plan Still Falls Short," April 26, 2010, p. 18 (http://www.spacenews.com/commentaries/100425-nasa-plan-falls-short.html).

U.S. Congress. "National Aeronautics and Space Administration Authorization Act of 2005," Public Law 109-155, sections 501-502, December 30, 2005.

White House Office of Management and Budget. "Budget for Fiscal Year 2011: National Aeronautics and Space Administration," February 1, 2010, pp. 129-132.

White House Office of Science & Technology Policy & NASA. "Fact Sheet: A Bold New Approach for Space Exploration and Discovery," February 1, 2010.

Wolf, Frank. "Don't Forsake U.S. Leadership in Space," *Space News*, April 26, 2010, p. 19 (http://www.spacenews.com/commentaries/100425-dont-forsake-leadership-space.html).

Chapter 5: Goodbye space shuttle, hello . . . what?

Aerospace Safety Advisory Panel. "Annual Report for 2011" (http://oiir.hq.nasa.gov/asap/documents/2011_ASAP_Annual_Report.pdf).

Alliant Techsystems Inc. "ATK and NASA Sign Space Act Agreement for Liberty Launch System," press release, September 13, 2011 (http://atk.mediaroom.com/index.php?s=118&item=1109).

Berger, Brian & Dan Leone. "NASA 2012 Budget Funds JWST, Halves Commercial Spaceflight," *Space News*, November 21, 2011, p. 10 (http://www.spacenews.com/civil/111118-nasa-budget-funds.html).

Booz Allen Hamilton. "Independent Cost Assessment of the Space Launch System, Multi-Purpose Crew Vehicle and 21st Century Ground Systems Programs: Executive Summary of Final Report," August 19, 2011 (http://www.nasa.gov/pdf/581582main_BAH_Executive_Summary.pdf).

Feinstein, Sen. Dianne, & Sen. Barbara Boxer. Letter to NASA Administrator Charles Bolden urging SLS booster competition, May 27, 2011 (http://www.spaceref.com/news/viewsr.html?pid=37310).

Griffin, Michael D. & Scott Pace. "Propellant Depots Instead of Heavy Lift?" *Space News*, October 31, 2011, p. 17 (http://www.spacenews.com/commentaries/111031-propellant-depots-instead-heavy-lift.html).

Hutchison, Sen. Kay Bailey, & Sen. Richard Shelby. "Failure of Russian Resupply Mission Underscores Need to Sustain America's Leadership in Space; Senators Again Call on NASA to Announce SLS Design Immediately," press release, August 24, 2011 (http://hutchison.senate.gov/?p=press_release&id=746).

Hutchison, Sen. Kay Bailey, & Sen. Bill Nelson. "Sen. Hutchison, Sen. Nelson Issue Statement on Campaign to Undermine America's Manned Space Program," September 8, 2011 (http://hutchison.senate.gov/?p=press_release&id=758).

Leone, Dan. "NASA Budget Bill Goes to House Floor Without Money for Webb; Support for Commercial Crew Transports Also Singled Out for Less Funding in 2012," *Space News*, July 18, 2011, p. 6 (http://www.spacenews.com/policy/110718-commercial-crew-transports-less-funding.html).

Leone, Dan. "Obama Administration Accused of Sabotaging Space Launch System," *Space News*, September 12, 2011, p. 1 (http://www.spacenews.com/policy/110909-obama-admin-accused-sabotaging-sls.html).

NASA. "Preliminary Report Regarding NASA's Space Launch System and Multi-Purpose Crew Vehicle," January 2011 (http://commerce.senate.gov/public/?a=Files.Serve&File_id=6bb9bc53-1ac8-457a-a5a2-018cbb8df292).

NASA. "ESD [Exploration Systems Development] Integration: Budget Availability Scenarios" briefing, August 19, 2011 (http://images.spaceref.com/news/2011/NASA.SLS.Budget.Aug.2011.pdf).

NASA. "NASA Announces Design for New Deep Space Exploration System: New Heavy-Lift Rocket Will Take Humans Far Beyond Earth," press release 11-301, September 14, 2011.

National Aeronautics and Space Act, Sec. 20102(d) (http://www.nasa.gov/offices/ogc/about/space_act1.html).

Nelson, Sen. Bill, & Sen. Marco Rubio. Letter to President Obama endorsing application of SLS funds to Kennedy Space Center projects, August 26, 2011 (http://spacepolicyonline.com/pages/images/stories/Nelson_Rubio_Ltr_SLS_Aug_26.pdf).

Olson, Rep. Pete, Rep. Bill Posey, Rep. Jason Chaffetz, Rep. Sandy Adams, Rep. Rob Bishop, & Rep. Mo Brooks. Letter to House Appropriations chairman Rep. Harold Rogers and House Commerce, Justice, & Science Subcommittee chairman Frank Wolf on reprogramming of NASA funds, February 7, 2011.

Rohrabacher, Rep. Dana. "Rohrabacher Reacts to Russian Soyuz Launch Failure; Calls for Emergency Funding of Commercial Crew Systems," press release, August 24, 2011 (http://rohrabacher.house.gov/News/DocumentSingle.aspx?DocumentID=257281).

Rohrabacher, Rep. Dana. "Fueling Stations vs. Monster Rocket," *Space News*, October 24, 2011, p. 19 (http://www.spacenews.com/commentaries/111024-fueling-stations-rocket.html).

Shelby, Sen. Richard. Letter to NASA Administrator Charles Bolden urging SLS booster competition, June 10, 2011 (http://images.spaceref.com/news/2011/20110610ShelbySLSBoosterLetter.pdf).

Shelby, Sen. Richard. Letter to President Obama alleging misapplication of SLS funds to Kennedy Space Center projects, August 15, 2011; co-signed by Senators Jeff Sessions, Thad Cochran, David Vitter, and Roger Wicker.

Space News editorial. "Soyuz Failure Highlights Priorities, Real and Political," September 5, 2011, p. 18 (http://www.spacenews.com/commentaries/110906-soyuz-failure-highlights-priorities.html).

Svitak, Amy. "NASA Sends Congress a Heavy-lift Design Too Big for its Budget," *Space News*, January 17, 2011, p. 5 (http://www.spacenews.com/civil/110111-nasa-heavy-lift-proposal.html).

Svitak, Amy. "GOP Lawmakers Appeal for Manned Exploration Funds," *Space News*, March 21, 2011, p. 3 (http://www.spacenews.com/civil/110317gop-lawmakers-appeal-for-manned-exploration-funds.html).

Senate Committee on Commerce, Science, and Transportation. Letter to NASA Administrator Charles Bolden regarding NASA Authorization Act of 2010 compliance, signed by Senators John D. Rockefeller (D-W. Va.), Kay Bailey Hutchison (R-Texas), Bill Nelson (D-Fla.), & John Boozman (R-Ark.), May 18, 2011.

U.S. Congress. "National Aeronautics and Space Administration Authorization Act of 2010," Public Law 111-267, October 11, 2010.

Chapter 6: Transition, or more of the same?

Achenbach, Joel. "Gingrich's moon shot: New Space Age or science fiction?" *Washington Post*, January 27, 2012, p. A6 (http://www.washingtonpost.com/national/health-science/newt-gingrichs-plan-for-a-moon-base-is-it-science-fiction/2012/01/26/gIQAKVC2TQ_story.html?sub=AR).

Armstrong, Neil, Eugene Cernan, & James Lovell. Letter to Rep. Frank Wolf, chairman of the House Appropriations Subcommittee on Commerce, Justice, Science, & Related Agencies, May 4, 2012 (http://www.americaspace.org/wp-content/uploads/space_docs/Armstrong_Cernan_Lovell_May_7_2012_Letter.pdf).

Bailey, Holly. "Long, volatile race tests Mitt Romney's enthusiasm: 'It's a very grueling process,' aide says," Yahoo news, February 27, 2012 (http://news.yahoo.com/blogs/ticket/long-race-tests-mitt-romney-enthusiasm-very-grueling-153955090.html).

Federal Aviation Administration, Office of Commercial Space Transportation. "Commercial Space Transportation: 2011 Year in Review," January 2012 (http://www.faa.gov/about/office_org/headquarters_offices/ast/media/2012_YearinReview.pdf).

Frank, Rep. Barney. "Mars Debate: Should U.S. Green-Light Manned Mission To Red Planet?" *Huffington Post*, April 2, 2012 (http://www.huffingtonpost.com/2012/04/02/mars-debate-us-manned-mission_n_1396949.html).

Fries, Sylvia D. "Opinion polls and the U.S. Civil Space Program," paper presented to the American Institute for Aeronautics and Astronautics, April 29, 1992, by the NASA Office of Special Services.

Grey, Jerry. *Enterprise*. (New York: William Morrow & Co., 1979).

Hutchison, Sen. Kay Bailey. "Senator Hutchison to Work with Colleagues to Restore NASA Human Exploration Funding," press release, February 13, 2012 (http://www.hutchison.senate.gov/?p=press_release&id=975).

King, Ledyard. "NASA leader defends budget," *Florida Today*, March 8, 2012 (http://www.floridatoday.com/article/20120308/NEWS02/303080051/NASA-leader-defends-budget?odyssey=tab%7Ctopnews%7Ctext%7CSpace&nclick_check=1).

Klotz, Irene. "Candidate Gingrich Promises Space Coast Voters the Moon," *Space News*, January 30, 2012, p. 5 (http://www.spacenews.com/civil/120126-gingrich-promises-space-coast-voters-the-moon.html).

Launius, Roger, & Howard McCurdy (eds.). *Spaceflight and the Myth of Presidential Leadership* (University of Illinois Press, 1997).

Leone, Dan. "Commercial Crew Backers Outline Budget Shortfall Survival Strategy," *Space News*, February 27, 2012, p. 14 (http://www.spacenews.com/civil/120227-commercial-backers-outline-budget-survival.html).

Leone, Dan. "Bolden, Lawmakers Lock Horns over Commercial Crew, Orion," *Space News*, March 12, 2012, p. 6 (http://www.spacenews.com/civil/120307-bolden-hutchison-commercial-orion-funding.html).

Logsdon, John M. "The Decision to Develop the Space Shuttle," *Space Policy*, Vol. 2, No. 2, 1986, pp. 103-119.

Morring, Frank Jr. "Dead On Arrival: NASA managers face another uphill battle on human spaceflight, science priorities," *Aviation Week & Space Technology*, March 12, 2012, p. 36 (http://www.aviationweek.com:80/publication/awst/loggedin/AvnowStoryDisplay.do?fromChannel=awst&pubKey=awst&channel=awst&issueDate=2012-03-12&story=xml/awst_xml/2012/03/12/AW_03_12_2012_p36-434276.xml&headline=Lawmakers+Unhappy+With+NASA+Budget+Priorities).

NASA Fiscal Year 2013 Budget Request Executive Summary (http://www.nasa.gov/pdf/622653main_b%20FY13_NASA_Budget_Estimates_Agency_Summary.pdf).

NASA Fiscal Year 2013 Budget Request Summary and Tables (http://www.nasa.gov/pdf/622654main_a%20FY13_NASA_Budget_Summary_and_Tables.pdf).

NASA Fiscal Year 2013 Budget Request—Exploration (http://www.nasa.gov/pdf/622972main_13-16%20Exploration%200214%20BW.pdf).

NASA Fiscal Year 2013 Budget Request—Planetary Science (http://www.nasa.gov/pdf/622650main_e%20FY13_NASA_Budget_Planetary.pdf).

Obama campaign fact sheet, "President Obama's Accomplishments for NASA and Florida's Space Coast," May 22, 2012 (http://spaceref.com/news/viewpr.html?pid=37135).

Obama campaign press release, "Will Mitt Romney Fire Space Advisor Michael Griffin for Proposing Permanent Moon Base?" May 24, 2012 (http://twitpic.com/9osupo/full).

Olson, Rep. Pete. "Colleagues Urge White House To Correct Safety Glitch," press release, February 29, 2012 (http://olson.house.gov/index.cfm?sectionid=129&sectiontree=21,129&itemid=945).

Powers, Scott. "Boisterous crowds bring energy to Mitt Romney campaign in Central Florida," *Orlando Sentinel*, January 27, 2012 (http://www.orlandosentinel.com/news/politics/os-romney-in-central-florida-20120127,0,6986039.story).

Romney Space Policy Advisory Group. "Romney Will Restore America's Space Program," Mitt Romney campaign press release, January 27, 2012 (http://mittromney.com/news/press/2012/01/leaders-americas-space-program-write-open-letter-support-mitt-romney).

Schiff, Rep. Adam. "Rep. Schiff Statement on Meeting with NASA Administrator," press release, February 9, 2012 (http://schiff.house.gov/latest-news/rep-schiff-statement-on-meeting-with-nasa-administrator/).

Space News. "A Bad News Budget Request," editorial, February 27, 2012, p. 18 (http://www.spacenews.com/commentaries/120227-bad-news-budget-request.html).

York, Byron. "In NASA-land, Romney, Gingrich battle over space," *Washington Examiner*, January 29, 2012 (http://campaign2012.washingtonexaminer.com/article/nasa-land-romney-gingrich-battle-over-space/346161).

Chapter 7: Planes, trains, automobiles, and spaceships

American Association of Port Authorities. "U.S. Public Port Facts" (http://www.aapa-ports.org/Industry/content.cfm?ItemNumber=1032&navItemNumber=1034).

Bilstein, Roger. *Orders of Magnitude: A History of the NACA and NASA*, 1915-1990 (Washington: NASA History Series SP-4406, 1989) (http://history.nasa.gov/SP-4406/contents.html).

Boyne, Walter J. *The Influence of Air Power upon History* (Gretna, LA: Pelican Publishing Company, 2003).

Cardwell, Donald. *The Norton History of Technology* (New York: W.W. Norton & Company, 1995).

Cruise Lines International Association (http://www2.cruising.org/industry/tech-intro.cfm).

Edwards, Chris. "A jobs plan we shouldn't bank on," *Washington Post*, October 23, 2011, p. B1 (http://www.washingtonpost.com/opinions/infrastructure-projects-to-fix-the-economy-dont-bank-on-it/2011/10/18/gIQAgtZi3L_story.html).

Hansen, Mark, Carolyn McAndrews, & Emily Berkeley. "History of Aviation Safety Oversight in the United States," U.S. Department of Transportation/Federal Aviation Administration, DoT/FAA/AR-08/39, July 2008.

Heppenheimer, T.A. *Turbulent Skies: The History of Commercial Aviation* (New York: John Wiley & Sons, 1995).

Holanda, Ray. *A History of Aviation Safety* (Bloomington, IN: AuthorHouse, 2009).

Levinson, Marc. *The Box: How the Shipping Container Made the World Smaller and the World Economy Bigger* (Princeton University Press, 2006).

Mazlish, Bruce. "Historical Analogy: The Railroad and the Space Program and Their Impact on Society" in Mazlish (ed.), *The Railroad and the Space Program: An Exploration in Historical Analogy* (MIT Press, Cambridge, Massachusetts, 1965).

McDonald, Charles W. "The Federal Railroad Safety Program," Federal Railroad Administration, August 1993 (http://www.fra.dot.gov/downloads/safety/rail_safety_program_booklet_v2.pdf).

Podkul, Cezary. "Behind the curve: With U.S. infrastructure aging, public funds scant, more projects going private," *Washington Post*, October 23, 2011, p. G1 (http://www.washingtonpost.com/business/with-us-infrastructure-aging-public-funds-scant-more-projects-going-private/2011/10/17/gIQAGTuv4L_story.html).

Richardson, Rickee. "Cruise Ship Safety," CruiseReport.com, not dated (http://www.cruisereport.com/NewsReader.aspx?news=453).

Roland, Alex. *Model Research* (Washington: NASA History Series SP-4103, 1985) (http://history.nasa.gov/SP-4103/contents.htm).

Swift, Earl. *The Big Roads: The Untold Story of the Engineers, Visionaries, and Trailblazers Who Created the American Superhighways* (Boston: Houghton Mifflin Harcourt, 2011).

Voulgaris, Barbara. "From Steamboat Inspection Service to U.S. Coast Guard: Marine Safety in the United States from 1838 1946," U.S. Coast Guard,

2009 (http://www.uscg.mil/history/articles/VoulgarisMarineSafety2009. pdf).

University of Wyoming, American Heritage Center. "Inventory of the Manufacturers Aircraft Association records," 2007 (http://ahc.uwyo.edu/ usearchives/inventories/html/wyu-ah06858.html).

U.S. Army. Letter contract for Wright B Flyer (http://www.wright-b-flyer. org/contract.html).

U.S. Centennial of Flight Commission. "The Wright Patent Battles" (http:// www.centennialofflight.gov/essay/Wright_Bros/Patent_Battles/WR12. htm).

U.S. Congress. Hours of Service Act, 45 U.S.C. 61 et seq., passed in 1907.

U.S. Congress. Safety Appliance Act, 45 U.S.C. 1-16, passed in 1893 and expanded in 1903 and 1910.

U.S. Congressional Budget Office. "U.S. Shipping and Shipbuilding: Trends and Policy Choices," August 1984 (http://www.cbo.gov/ftpdocs/113xx/ doc11317/1984_09_shipping.pdf).

Williamson, John. "Federal Aid to Roads and Highways Since the 18th Century: A Legislative History," Congressional Research Service report R42140, January 6, 2012 (http://www.fas.org/sgp/crs/misc/R42140.pdf).

Wolmar, Christian. *Blood, Iron, & Gold: How the Railroads Transformed the World* (New York: Public Affairs, 2010).

Chapter 8: Finding a path to the mainstream

David, Leonard. "Is Mining Rare Minerals on the Moon Vital to National Security?" *Space.com*, October 4, 2010 (http://www.space.com/news/moon-mining-rare-elements-security-101004.html?utm_source=feedburner&utm_medium=feed&utm_campaign=Feed%3A+spaceheadlines+%28SPACE. com+Headline+Feed%29&utm_content=My+Yahoo).

Kennedy, John F. Memo to Vice President Lyndon Johnson, April 20, 1961 (http://www.jfklibrary.org/Asset-Viewer/6XnAYXEkkkSMLfp7ic_o-Q. aspx).

Kennedy, John F. Special Message to Congress on Urgent National Needs, May 25, 1961 (http://www.jfklibrary.org/Research/Ready-Reference/ JFK-Speeches/Special-Message-to-the-Congress-on-Urgent-National-Needs-May-25-1961.aspx).

Kennedy, John F. Address at Rice University on the Nation's Space Effort, September 12, 1962 (http://www.jfklibrary.org/Research/

Ready-Reference/JFK-Speeches/Address-at-Rice-University-on-the-Nations-Space-Effort-September-12-1962.aspx).

Letter to Senators Barbara Mikulski and Kay Bailey Hutchison, chair and ranking member of the Senate Appropriations Subcommittee on Commerce, Justice, and Science, on fiscal year 2012 funding for NASA's Space Technology Program, signed by 45 companies and universities, September 8, 2011 (http://commercialspaceflight.org/Other%20Content/NASA%20Space%20Technology%20Letter%20of%20Support,%20Senate.pdf).

Logsdon, John. *John F. Kennedy and the Race to the Moon* (New York: Palgrave Macmillan, 2010).

Robinson, Michael & Dan Lester. "NASA and the Ghosts of Explorers Past," *Space News*, February 8, 2010, p. 17 (http://www.spacenews.com/commentaries/100208-nasa-ghosts-explorers-past.html).

Chapter 9: In search of forward-looking space policy

Bush, George W. "U.S. Space Exploration Policy," January 14, 2004 (http://www.fas.org/irp/offdocs/nspd/nspd-31.pdf).

Heclo, Hugh. "Issue Networks and the Executive Establishment" in A. King (ed.), *The New American Political System* (Washington: American Enterprise Institute, 1978).

Marburger, John. "Remarks on the background for the Vision for Space Exploration presented to the Review of U.S. Human Space Flight Plans Committee," August 5, 2009 (http://www.nasa.gov/pdf/376646main_11%20-%20jhm%20Augustine%20Committee%2009-05-09B.pdf).

National Commission on Space. *Pioneering the Space Frontier* (New York: Bantam Books, 1986) (http://history.nasa.gov/painerep/begin.html).

Noordung, Hermann. *The Problem of Space Travel: The Rocket Motor* (Washington: NASA History Series SP-4026, 1995). English translation of a book originally published in German in 1929.

Obama, Barack. "National Space Policy of the United States of America," June 28, 2010 (http://www.whitehouse.gov/sites/default/files/national_space_policy_6-28-10.pdf).

Obama, Barack. "Remarks by the President on Space Exploration in the 21st Century," Kennedy Space Center, Florida, April 15, 2010 (http://www.whitehouse.gov/the-press-office/remarks-president-space-exploration-21st-century).

President's Commission on Implementation of United States Space Exploration Policy (Aldridge Commission). "A Journey to Inspire, Innovate, and Discover," U.S. Government Printing Office, June 2004 (http://www.nasa.gov/pdf/60736main_M2M_report_small.pdf).

Space Task Group, "The Post-Apollo Program: Directions for the Future," September 1969 (http://www.hq.nasa.gov/office/pao/History/taskgrp.html).

Chapter 10: The Next Great Thing

Castronuovo, M.M. "Active space debris removal—A preliminary mission analysis and design," *Acta Astronautica*, April 2011 (http://www.sciencedirect.com/science/article/pii/S0094576511001287).

DeSelding, Peter B. "Intelsat Signs Up for Satellite Refueling Service," *Space News*, March 14, 2011 (http://www.spacenews.com/satellite_telecom/intelsat-signs-for-satellite-refueling-service.html).

General Accounting Office. "NASA Has No Firm Need for Increasingly Costly Orbital Maneuvering Vehicle," NSIAD-90-192, July 31, 1990 (http://www.gao.gov/assets/150/149382.pdf).

Horsham, Gary A.P., George R. Schmidt, & James H. Gilland. "Establishing a Robotic, LEO-to-GEO Satellite Servicing Infrastructure as an Economic Foundation for Exploration," AIAA 2010-8897, presented at AIAA Space 2010, August 30-September 2, 2010.

Jet Propulsion Laboratory. "NASA in Final Preparations for Nov. 8 Asteroid Flyby," October 26, 2011 (http://www.jpl.nasa.gov/news/news.cfm?release=2011-332).

Klamper, Amy. "Bigelow Modules Draw Interest from Six Governments," *Space News*, October 25, 2010, p. 12 (http://www.spacenews.com/venture_space/101022-bigelow-modules-interest-six-governments.html).

Lewis, Jeffrey. "Autonomous Proximity Operations: A Coming Collision in Orbit?" University of Maryland, March 11, 2004 (http://www.cissm.umd.edu/papers/files/autonomous_proximity.pdf).

Liou, J.C. "An active debris removal parametric study for LEO environment remediation," *Advances in Space Research*, Vol. 47 (2011), pp. 1865-1876 (http://www.sciencedirect.com/science/article/pii/S0273117711000974).

Moltz, James Clay. *The Politics of Space Security* (Stanford, CA: Stanford University Press, 2008).

United Nations. Convention on Registration of Objects Launched into Outer Space (Registration Convention) (http://www.oosa.unvienna.org/oosa/

en/SORegister/regist.html). Entered into force in 1975. Ratified by 56 countries and signed by four as of January 2011.

United Nations. Treaty on Principles Governing the Activities of States in the Exploration and Use of Outer Space, including the Moon and Other Celestial Bodies (Outer Space Treaty), January 27, 1967 (http://www.oosa.unvienna.org/oosa/en/SpaceLaw/outerspt.html). Ratified by 101 countries and signed by 26, including all major spacefaring nations, as of January 2011.

U.S. Government Orbital Debris Mitigation Standard Practices, December 2000 (http://www.iadc-online.org/index.cgi?item=documents).

Werner, Debra. "NASA Defends On-orbit Satellite Refueling Demonstration," *Space News*, June 27, 2011, p. 10.

Yeomans, Don, Lance Benner, & Jon Giorgini. "Asteroid 2005 YU55 to Approach Earth on November 8, 2011," Jet Propulsion Laboratory, March 10, 2011 (http://neo.jpl.nasa.gov/news/news171.html).

Chapter 11: Answering the big questions

International Space Exploration Coordination Group. "The Global Exploration Roadmap," published by NASA Headquarters (NP-2011-09-766-HQ), September 2011 (http://www.globalspaceexploration.org).

Leone, Dan, "NASA Completes 1st Robotic Satellite Servicing Demonstration at ISS," *Space News*, March 19, 2012, p. 7 (http://www.spacenews.com/civil/nasa-completes-first-robotic-satellite-servicing-demonstration-iss.html).

Mankins, John. "Technology Readiness Levels: A White Paper," NASA Advanced Concepts Office, April 6, 1995 (http://www.hq.nasa.gov/office/codeq/trl/trl.pdf).

NASA Draft Space Technology Roadmaps, November 2010 (http://www.nasa.gov/offices/oct/home/roadmaps/index.html).

NASA Public Affairs, "NASA and CSA Robotic Operations Advance Satellite Servicing," Press Release 12-080, March 13, 2012 (http://www.nasa.gov/home/hqnews/2012/mar/HQ_12-080_Robotic_Refueling_Mission.html).

NASA Public Affairs, "NASA Survey Counts Potentially Hazardous Asteroids," Press Release 12-157, May 16, 2012 (http://www.nasa.gov/home/hqnews/2012/may/HQ_12-157_NEOWISE_PHAs.html).

National Research Council. "An Interim Report on NASA's Technology Roadmaps," 2011 (http://www.nap.edu/catalog.php?record_id=13228).

National Research Council. "NASA Space Technology Roadmaps and Priorities: Restoring NASA's Technological Edge and Paving the Way for a New Era in Space," 2012 (http://www.nap.edu/catalog.php?record_id=13354).

Chapter 12: Not the end

NASA Advisory Council. Letter to NASA Administrator Charles Bolden signed by Council Chairman Steven Squyres, March 20, 2012 (http://www.nasa.gov/pdf/633510main_12-03_recommendations.pdf).

NASA Innovative Advanced Concepts Program (http://www.nasa.gov/offices/oct/early_stage_innovation/niac/index.html).

NASA Institute for Advanced Concepts, studies performed 1998-2007 (http://www.niac.usra.edu/studies/studies.html).

www.ingramcontent.com/pod-product-compliance
Lightning Source LLC
Chambersburg PA
CBHW030930180526
45163CB00002B/519